PRAISE FOR

The Voyage of the Yacht, *Dal*

I am in awe of Irene Tomaszewski's introduction and translation of the story of the *Dal* and her portrait of Andrzej Bohomolec, and am grateful for the opportunity to be one of the first to read it. Personally, I feel a connection to Andrzej Bohomolec—as a dreamer, in his need to organize such a bold and risky voyage, and living through the experience, surviving the catastrophe, and finding such a success... *Dal* is one of the most beautiful words in Polish, and yet so hard to translate into English. In a way, the *Dal* is a symbol of immigrants... an independence gained by decisions, action, and being alone in a dream.

Małgorzata Kot, MA, MLIS
Managing Director,
The Polish Museum of America, Chicago

This is a book in two parts. One is about an exciting adventure by three young men feeling free in their own country. The other is about one of these men a few years later during WWII. He lost everything after fighting against both the Nazis and the Soviets and couldn't even go home. In Canada he found a country he loved and admired. It's a unique story, but thousands of Polish Canadians can relate to it.

John Tomczak
President
Canadian Polish Congress

Fascinating story from an adventurer who embodied all that is to be a patriot, warrior, philosopher, artist. In other words a true renaissance man. He reminds us that you should search for freedom in the challenges that only nature can bring, in the mountains and in the sea.

Antoni Kowalczewski
President
Poland in the Rockies Society.

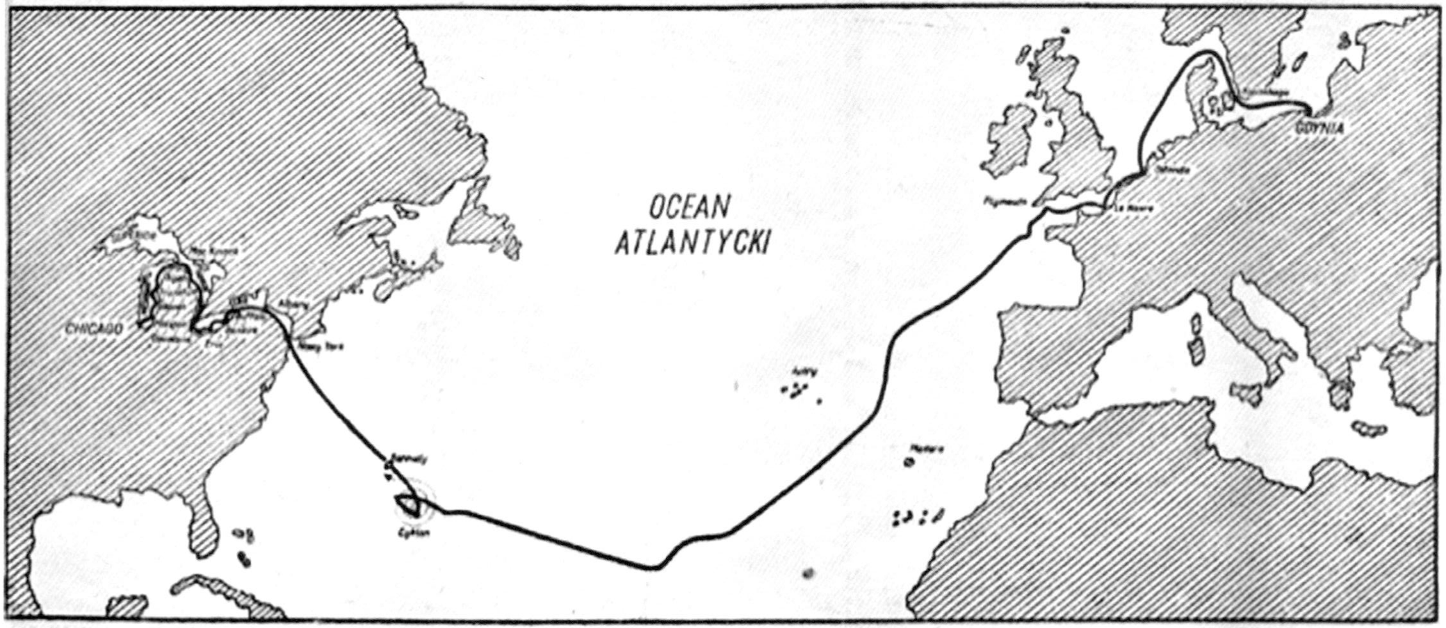

TRASA PODRÓŻY JACHTU „DAL"

Gdynia–Copenhagen–Ostend–Le Havre–Plymouth–Bermuda–New York–West Point–Albany–Schenectady–Amsterdam–Utica–Breverton–Rochester–Buffalo–Cleveland–Hamtramck–Alpena–Mackinac–Ludington–Grand Haven–Michigan City–Chicago.

The Voyage of the Yacht, *Dal*

from Gdynia to Chicago, 1933–34

A. Bohomolec

Translation and Introduction

by

Irene Tomaszewski

Published by: The Canadian Foundation for Polish Studies
of the Polish Institute of Arts and Sciences in Canada, Montreal-Ottawa
Contact: cfps.ottawa@gmail.com

1st edition: WYPRAWA JACHTU "DAL"
published by Towarzystwa Wydawnicze "RÓJ" 1936

2nd edition: WYPRAWA JACHTU "DAL"
published by Wiedza Powszechna 1957

Author: André (Andrzej) Bohomolec
Translation and Introduction by Irene Tomaszewski
The Voyage of the Yacht, *Dal*: from Gdynia to Chicago, 1933–34

Cover illustration and line drawings illustrating text are by Adam Werka.

Issued in print and electronic formats
Book: ISBN 978-0-9868851-7-4
E-book: ISBN 978-0-9868851-8-1

Cover and Book Design: Budd Publishing, Ottawa, Ontario
Set in: Warnock Pro and Chaparral Pro

Printer: Gilmore Printing Services, Ottawa, Ontario

PRINTED IN CANADA

1957 edition

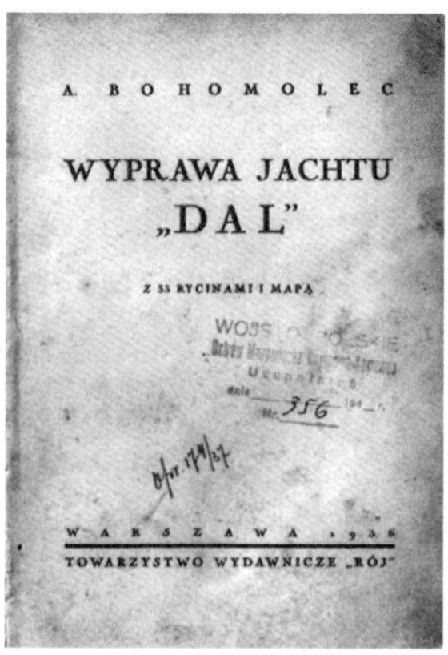

1936 edition

MIX
Paper from
responsible sources
FSC® C100205

Andrzej (André) Bohomolec, ca 1951

Table of Contents

Foreword

In the 1950s in Alberta I learned from my stepfather André Bohomolec that he and two companions had made a voyage from Poland to Chicago in the 1930s for the World's Fair, that they were almost wrecked in a hurricane and repaired the boat in Bermuda. Later, André gave me a copy of his 1957 book describing the voyage, but none of my family reads Polish.

Recently, I decided that the book should be translated into English. But who could do it? A chance encounter with Paweł Wołowski in Ottawa was the key. Paweł most kindly offered to try to find a writer who would do the translation. By great good fortune, Irene Tomaszewski agreed.

After a quick read, Irene telephoned me to say that the book's introduction said nothing about the author, and that she had learned that the book was well known in the world of Polish yachting. So she proposed to compose a new introduction describing the author. Irene, assisted by her friend Małgorzata Dzieduszycka in Warsaw, has carried out an astonishing piece of research. Not only do we now know of André's life before he came to Canada, but we are introduced to his Polish family and its background in the context of history. And his story of a remarkable voyage is now available to the English speaking world in a beautifully fluent translation. We are all in Irene's debt.

Anthony Keith
Ottawa
March 2019

Introduction

While rather reluctant to read, let alone translate, a book about sailing—about which I know nothing—I agreed to "just read it" as a courtesy to the person who asked me, Paweł Wołowski, to whom I am grateful for his persuasiveness. A quick reading charmed me and immediately left me wondering about the author: A. Bohomolec.

Who was he? Andrzej (André) Bohomolec, soldier, sailor, writer, diplomat, commando, liaison officer, artist, rancher, and an exile who found in multicultural Canada the ideal country.

The book recounted an engaging story about a 1933 trans-Atlantic crossing in an 8.5 metre boat not quite fit for such an undertaking. But it was the best the sailors could afford. Their daring "can-do" spirit, brilliant sailing and navigating skills, and the understated literary style of the mysterious author intrigued me. And then there was a certain air of mystery about the book itself.

The publishing date is not usually the first thing to be noticed but when a book is published 23 years after the event, in Communist Poland, then the curiosity of a reader who knows something about Poland is sparked. Why in 1957, and why not a word about the author? Had there been an earlier edition and if so why is there no reference to it?

I plunged into the translation, and at the same time began an investigation. Why did this man and this book suddenly appear in 1957 Poland? The project was made easier by collaborating with partners in Warsaw. Małgorzata Dzieduszycka, an author, literary critic, former diplomat and a dear friend, stepped into the spirit of the mystery and also checked my translation. She found an expert to help with the technical terms of the seas, currents, winds and the technique of sailing. Captain Krzysztof Baranowski, yachtsman, journalist, teacher, and a member of the Council on the Board of the Polish Maritime Foundation, he was the first Pole to sail around the globe, twice. He also clarified one thing immediately: every Polish sailor knows about Andrzej Bohomolec.

There was little doubt in my mind that there had been an earlier edition, and soon some sleuthing by Małgorzata cleared that up. She vaguely remembered seeing the 1957 edition in her father's library and agreed that there must have been an earlier edition. Searching antiquarian bookshops, she struck gold: a hard cover edition of the original 1936 book, complete with photographs of the mariners. The author, it turned out, was as handsome as he was daring. The 1957 edition had no photographs but was illustrated with lovely pen and ink drawings.

The average western reader knows little about Soviet-style Communism, and some entertain fanciful notions about it. Revelations about its brutality were long delayed, but some things about the sheer lunacy of its policies may not have reached even the sophisticated reader.

Can we imagine sailing being banned because it was a bourgeois sport? Can we understand a government that would confine watercraft to inland lakes and rivers because sailing on the wide open sea gives one an exhilarating sense of freedom that could inspire thoughts of escape? Winston Churchill's famous "Iron Curtain" descended on Europe, but how does one install that on an ocean? Where would the checkpoints be?

A quick chronology of events is in order. The voyage of the little yacht, the *Dal*, took place just 13 years after Poland had thrown off its imperial rulers, Russia, Prussia and Austria. Of the three, Russia was the least willing to accept this outcome so in 1919-20 the Poles had to defend their country's newly won independence. The author, then just 19 years old, fought in that battle. The country was secured, complete with a coastline. Not a long one, but a coastline nevertheless. After 125 years of foreign rule and once again a sovereign country, Poland built a modern seaport at Gdynia, sailors established a yacht club, and the government even encouraged the sport with organizations of sea cadets. The young people of Poland flocked to the sea.

The first edition of André's book was a bestseller. Every schoolboy wanted a copy and dreamed about adventures until recently closed to them. In 1939 their dreams were shattered. Once again those empires, this time under different names, attacked and occupied the country. The Soviets took the eastern part, arresting, deporting and killing to cement their conquest. The Germans took the western part. In 1941, they turned on their erstwhile ally, and occupied all of Poland for a total of six long and murderous years.

Apart from the cruelty towards the people, the German occupying forces practiced a policy of cultural destruction, with libraries a principal target. Libraries were not just looted and bombed, the Germans used flamethrowers on great libraries to ensure nothing would survive the destruction of the building. André's family home in Warsaw was bombed and burnt to the ground during the 1939 siege of Warsaw. If there were copies of the book there, they perished. The Polish capital, Warsaw, had been levelled to the ground, its population expelled.

In 1945 the war ended in the western countries of Europe but Poland was once again occupied. Soviet Russia, while defeating its former ally, occupied all of Poland and imposed a Communist regime on the devastated country. André, along with all other Poles who had fought with the Western allies, could not return home.

As the country gradually recovered, housing, schools, hospitals and other public institutions, among them libraries, were slowly rebuilt. Professors who had been teaching from memory were writing new books, publishing houses were established. Still, the Communist regime was controlled by Moscow, as was publishing, and books were carefully censored. Even a new translation of Herodotus was delayed—who knows why? Perhaps because the ancient Greek wrote about his travels and that could make people think about far away places and different cultures. The censorship business does not attract the best minds.

Can sports be censored? The Communist censors thought so: sailing was not the sport of the proletariat and André Bohomolec was not a proletarian. But more about that later.

In 1956, the communist regime in Poland went through a period known as "The Thaw," a loosening of Moscow's rules. This was largely due to the welcomed death of Stalin who finally relieved the people of his rule in 1953. And so, thousands of political prisoners were released, death sentences were commuted, Herodotus—and even *The Voyage of the Yacht,* Dal—was published.

Once again the book was a bestseller, and along with the book was a lift of the ban on sailing. The schoolchildren who had read André's book in 1936, now adults, had not forgotten their hero. Sailing clubs sprang up everywhere, Poles started sailing in international regattas and with Tall Ships and, in 1976, Krystyna Chojnowska was the first woman to sail solo around the world. And André Bohomolec remains, to this day, an iconic figure in sailing, the 80th anniversary of his voyage commemorated at a special event at the Gdynia Public Library.

So, who was he?

André Bohomolec was born on November 22, 1900, in Poland, on the family estate in Rozentowo, Poland. Or, as the French writer, Alfred Jarry, would have put it, he was born "in Poland, that is to say, nowhere." In 1900, Rozentowo was in a province of the Russian empire, but over the centuries it had been in a principality of Minsk, in the Grand Duchy of Lithuania, in the Polish-Lithuanian Commonwealth, in Belarus, in the Latvian Socialist Republic, and in Latvia. In 1958 when André's Canadian citizenship certificate was issued it states that he was born in "Rozentowo, Poland," even though he knew that the house where he was born and the church where he had been baptized were gone, bombed during the First World War, and burnt after the Russian Revolution. Unlike *Monsieur* Jarry, André never had any doubts about the country where Rozentowo was; *Monsieur* Jarry was quite wrong: far from being "nowhere," Poland was everywhere. And whatever the borders imposed by Russia, he was raised in the tradition of the Polish-Lithuanian Commonwealth, a vast territory bound by two seas, the Baltic and the Black, a country formed by a union of two nations with two official languages, comprising people of many ethnicities, professing a variety of religions, practicing various traditions, speaking many languages.

He was the last of a family of 23 generations, first documented in 1098. The record of his princely family

Rozentowo

includes scholars and soldiers, clergy and entrepreneurs. André always wore his father's signet ring with the Bogoria coat-of-arms on it. The coat-of-arms in Poland, called a *herb*, had a different significance from that of the English or German which were granted to an individual or a family by a sovereign. The Polish crests, on the other hand, were accepted, rather than granted, and signified a brotherhood-in-arms with a large number of families freely united, obviously for mutual defence. The Bogoria crest and its holders date back to the 12th century. The young men from this group were sent to study at the Universities of Padua, Rome or Bologna, and a stone with this crest once embedded in a wall of a church can now be found in the municipal museum of Bologna. His family's origins, confirmed by Professor Oskar Halecki—an authority on the history of the Commonwealth whom André met at Columbia University in New York—meant a great deal to him.

In his book, André refers to one ancestor, Franciszek (1720-1784), with a modest apology for not measuring up to his literary standards. Franciszek had studied in Bologna, became a Jesuit only later in life after he had married, had several children, and was widowed. Not an unusual pattern at the time. (Rather sensible, really.) Franciszek was a noted author, translator (Latin and Italian), playwright, poet, essayist, and editor of a review, the *Monitor*. He is still studied today in Polish classical literature. On his mother's side he was also related to Henryk Sienkiewicz who won the Nobel Prize for literature in 1905.

The Jesuits were the main reason André's part of Poland remained Catholic during the Reformation when quite a few families converted to Lutheranism, notably in areas that are now Latvia and Estonia. The conversions were accepted as a matter of personal choice and the country was not involved in Europe's religious wars of the time, other than providing a refuge for dissenters from other countries. In southern Latvia, around Rozentowo, the area remained Catholic, largely due to the influence of the Jesuits. This Jesuit connection may well explain a humorous anecdote told by the daughter of one of his friends, some time after they all arrived in Canada. The girl attended a convent school and wanted to change to a public school. Thinking her "uncle André" would be her ally, she was surprised when he said, "Don't be silly. You are a Catholic, your family has always been Catholic, and so shall you remain." Startled, she said, "But Uncle André, you don't go to church!" To which he curtly replied. "That doesn't matter."

The estate in Rozentowo was a large enterprise producing superb cheeses that were sold in St. Petersburg and Warsaw, while nearby at Gorykolno, they grew flax for the linen producers who sold their product to tradesmen in Belgium. The Russian and German armies destroyed Rozentowo and the Russian revolution confiscated what was left. His parents relocated to Poland where his father established a pulp and paper enterprise, employing quite a few refugees from the Russian revolution. It was in Włocławek, an industrial city not

Rozentowo Cheese ad in 1910

far from Warsaw. During the Second World War the company was first taken over by the Germans and in the end by the Communist government established by Moscow.

André's early education was at home with tutors, with an emphasis on languages. André mastered fluency in Polish, French and Russian. At age 12 or so, he was sent to a Russian boarding school outside St. Petersburg. Russia had banned Polish schools. The family frequently travelled to France and André studied briefly in Switzerland, long enough to acquire a basic knowledge of German.

When the revolution broke out in 1917, André left the country via Finland, then by boat to France to join an army of volunteers, mostly Polish Americans but also some Canadians. This army, under the command of General Józef Haller, had trained in Niagara-on-the-Lake in Canada, since the US, being neutral at the time, could not permit a foreign army in its territory. The Haller Army rejected André because he was too young so he joined the French Foreign Legion until his 18th birthday. By this time the army was in Poland defending Poland's newly regained freedom. André joined them, was wounded during the Polish-Soviet war, (he lost two fingers), and was promoted to 2nd lieutenant at war's end. It was then time to continue his education.

His university education was in French, first in Leuven, Belgium, and then in Paris at the Institute of Political Studies (the *Sciences Po*). Returning to Poland, he was not enthusiastic about a military career but was keen about sailing. It seems he was inspired by the writings of Joseph Conrad, another Pole, whose family was dispossessed by Russia in the 19th century. Conrad wrote about his travels with the British merchant marine to Indo-China and Africa, André went to work with the

French merchant vessel, the *Chargeur Réunis*, sailing to Indochina, Africa and South America.

Upon his return, he went back to the cavalry but couldn't get the sea out of his mind and wanted to sail across the Atlantic. The government was keen to encourage young Poles to take an interest in the navy so he obtained leave to pursue this goal. He met two young sailors at the yacht club in Gdynia, Jan Witkowski and Jerzy Świechowski. Younger and less experienced than André—although Świechowski was already a superb navigator—they had not yet sailed beyond the Baltic, and were keen to join him on this adventure.

The transoceanic adventure is best related by the sailor himself. For someone unfamiliar with sailing, I can only add that starkly, without embellishment, he conveyed the need to remain alert even during periods of tedium because the life threatening force of nature could strike almost without warning. Their encounter with a hurricane not only blew them off course but left them with a crippled craft: broken mast and knotted rigging. Świechowski, the navigator who liked to sing, usually off key, expertly guided them to Bermuda using his watch and a sextant.

There we see more of André's personality, a man at ease in any society. Interested in the history, including the social history, of the island, André explored the many little islands, learning about their formation, the creatures in their coastal waters, which he would later appreciate as a subject for his ceramic art, and their earlier inhabitants. He noted the social stratification, ranging from the British ruling class to the moneyed and temporary American residents, the fishermen, the tavern keepers, and the friendly but largely excluded descendants of African slaves. He was interested in, and comfortable with, all of them, whether as a guest dining with the Commodore of the Yacht Club, eating on the beach with the fishermen, or just being friendly with the neighbours around the rather primitive but decidedly exotic house he and Świechowski had rented for the winter. Witkowski, the third sailor, suffering from the effects of the hurricane, had returned to Poland. The island would remain a fond memory and play an important part in André's life again, after the second great European war.

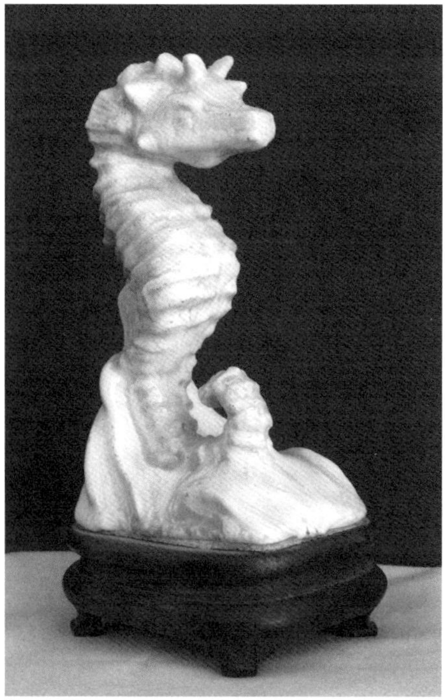

Ceramic seahorse

The next leg of their journey took them to New York. The two sailors were surprised to be met like celebrities, introduced to what Andy Warhol would some day label "fifteen minutes of fame."

Then, encouraged to carry on to Chicago, to the World's Fair, "The Century of Progress," they gamely took on the very different challenges of sailing inland rivers and canals, and the majestic Great Lakes. After more adventures and misadventures, they arrived at their destination, where André had to deal with more unexpected demands, speeches, interviews and, most unexpectedly, even a request to donate the *Dal* to the Polish American community and leave it at the

Exhibition. Reluctantly, but out of gratitude for his warm welcome, he agreed and while Świechowski went back to Poland, he went on a multicity tour to recoup some of the cost of his yacht.

After several months he finally returned home, wrote the book, published it in 1936, went back in the cavalry but was seconded to the Ministry of Foreign Affairs as an advisor on Soviet matters. In 1938 he was posted as Consul-General in Shanghai. Life was good.

Until 1939, when Germany attacked Poland on September 1, and Russia attacked from the east on September 17. Polish forces continued fighting until October 5 at which point the soldiers who were not taken as POWs went underground, others headed for France to regroup and continue the fight with French and British forces, but Poland did not surrender. André received these grim communiqués daily at his consulate in Shanghai.

The War Years

He decided to leave Shanghai and headed for France, arriving in time to join the 55,000 Poles fighting alongside the British and the French. Germany had already defeated Belgium, Luxembourg and the Netherlands. Within six weeks, France surrendered and formed a government in Vichy, subservient to and collaborating with the Germans in Paris.

André stayed in France until June 20, directing part of the evacuation in Operation Aerial, the final phase of evacuations to Britain from ports on the west coast of France following the earlier ones from Dunkirk and Le Havre. The last 9,000 Polish troops were evacuated along with diplomats, administration officials, and other civilians, among them women and children. The French writer and politician Maurice Schumann and Ronald Hugh Campbell, the British ambassador to France, joined this evacuation, picked up at Saint-Jean de Luz by two Gdynia America Line ships, the Batory and the Sobieski. André's record shows him as the Liaison Officer of the Polish Ministry of War and the last man on board.

He remained in the UK until Sept. 28, 1941, when he was posted as Liaison Officer of the Polish Forces with the British and Allied Forces in the Middle East in Cairo. Knowing others were fighting while he was attending meetings and social events did not please him so he requested permission to join the Polish Independent Carpathian Brigade. Officially established in April 1940 in Homs, Syria, the Brigade was formed by order of the Polish government's Commander-in-Chief, General Władysław Sikorski, who decided that a Polish unit should join the French in their territory of the Levant. When the French capitulated and the French commander in Syria decided to support the Vichy government, the Poles defected to the British in Palestine.

The Carpathian Brigade was an unusual, elite unit in which the men were older than the average in most military units, and 25 percent had university educations. They were volunteers committed to a cause. André fitted in well. Fighting alongside the British 8th Army and the 9th Australian Division, the Brigade distinguished itself at the Siege of Tobruk. Wounded in Bardia, André returned to England to convalesce but within a year he was back in North Africa, training with the SOE 9th Inter-Allied Commando Company.

In December 1943 he parachuted into northern Italy on a commando mission and connected with the Polish II Corps under the command of General Władysław Anders, then preparing for battle at Monte Cassino and opening the road to Rome. André was recalled to Headquarters General Staff in England in March 1944 and then assigned as Liaison Officer to the French Forces in anticipation of the liberation of Paris. His notes reveal nothing about his feelings witnessing this joyful occasion at a time when his countrymen were fighting the hopeless battle of the Warsaw Uprising... and Polish forces in the West were denied permission to go to their aid. The Polish Parachute brigade had trained specifically for that purpose but was ordered instead to take part in the foolhardy and suicidal battle at Arnhem in the Netherlands. While the Polish resistance fought against the full might of the German artillery, tanks and fighter planes unaided, their eastern "ally," the Soviet Union, camped on the other side of the Vistula River with orders from Stalin to advance no further. Stalin even refused landing and refuelling rights to allied volunteer airmen ferrying supplies to the Warsaw insurgents. Winston Churchill asked Roosevelt to join him in a request to Stalin to help Warsaw but Roosevelt did not want to antagonize "Uncle Joe," as he called him. Churchill's personal note to Stalin was ignored.

While Parisians celebrated their liberation, André was assigned the task of dealing with the thousands of liberated Polish prisoners roaming around in France. Although released from German camps, they had nowhere to go. For the time being, they were among many other released prisoners but the French, the Belgians, and the Dutch were able to go home as soon as the allied armies liberated their countries. The Polish First Armoured Division fighting alongside the First Canadian Army liberated one country after another. Poland had no such luck.

In this crush of former prisoners there were also some men from the Polish forces in Italy who, once hostilities had ended there, rushed to France in the hope of finding, or at least hearing about, family members.

By this time, the Polish Government-in-exile and the Polish Forces knew that Churchill and Roosevelt had already given in to Stalin's demands and ceded all the Polish territory—the eastern half of the country—he had seized in 1939. For most of them, there would be no going home even when the war ended. Despite the bitter experience of being unable to help his own country and people while fighting for the liberation of Western Europe and witnessing the joyful liberation of Paris, in his own notes André expresses no ill will to France and writes rather sympathetically about de Gaulle. After the war, as an avid reader of the Paris-based literary journal of the Polish exiles, *Kultura*, André was well aware that de Gaulle, alone among Western leaders, recognized the grave injustice inflicted on Poland. Among the editors of *Kultura*, were his friends, Polish veterans who suddenly became exiles; his sister, Adela, also worked with them. De Gaulle recognized *Kultura* as "the journal of an occupied country," and instructed his Minister of Information, André Malraux, to give it all the support it needed, especially in view of the vituperative attacks by the Communist press, notably the Stalinist *l'Humanité*.

But in the immediate aftermath of France's liberation, André and all his fellow officers were concerned with the situation of their soldiers, airmen, and sailors who, after fighting on every front had no victory to celebrate, and suddenly heard their allies referring to them as fascist enemies. They were even excluded from the victory parade in London. However, on a more personal front, the liberation of Paris did bring the Bohomolec family together again. While André continued his work with the soldiers, his sisters—Adela who also worked for the government-in-exile and Anna who worked for the International Red Cross—continued with their responsibilities helping the civilian Poles liberated from German camps. His mother, who remained in her apartment in Paris throughout the war, was there too. And then there was the French branch of the family in Paris, established in the early part of the 19th century.

André had to return to London to be demobilized, and honourably discharged. But he didn't linger in England, nor did he want to live in Paris, despite having family and friends there. They were all exiles. He preferred to get away from Europe, from war, from disillusionment, from bitter memories, and to art and contemplation. He chose Bermuda, an island of fond memories, unclouded by war and loss. The one thing that Polish veterans of forces under British command got after losing everything else was the status of "British subject," facilitating their entry into the Dominions and colonies still linked to Britain.

His official war record lists the medals he was awarded. From Poland: Cross of Valour (1919), Cross of Valour (1942), Gold Cross of Merit, Campaign Medal 1918/21 War, Bronze and Silver Medals for Long Service. From Britain: 1939/45 Star, Africa Star, Italy Star, France and Germany Star, Defence Medal 1939/45. From France: Chevalier de la Légion d'Honneur, "Étoile Noire", Medaille interallié, Medaille commemorative française de la guerre 1939/45.

Bermuda

In 1942, when he lay wounded in a tent hospital in North Africa, in pain and surrounded by death, he had experienced strange dreams during his feverish delirium. Later, convalescing in London, his thoughts returned to them, and to a philosophical contemplation of life. He wrote them down and gave them to his friend, Hereward T. Watlington, who translated them from the French and published them in a book, "Can an Armchair Fly?" in Bermuda in 1944. The author's name remained anonymous but Watlington writes, "... it is my hope that when the Author reads the printed page he will feel that his meaning and intention have been conveyed." There is also a dedication, "To my Mother and my Sisters in enemy occupied Territory," and the Translation's Foreword ends with, "Any profit which may accrue from the sale of this book will be devoted to the Welfare Section of the Polish Army."

His first thoughts while in the hospital, reflecting on pain, were: "soldiers, women bearing children, the Christ." He thought about life, the materialistic world and the mystical. "No longer a Churchman, I have a quantity of ideas often contradictory, which need to be put in order. But at this moment, I pray with all my heart to a Good God seated on a cloud and to the Holy Virgin in a robe of blue standing on a crescent... as I used to contemplate them each Sunday as a small child..." He wrote that children have a natural affinity for the supernatural and recalled a childhood conversation with his father who tried to teach him a materialist view of the world. Little André disagreed. His father said, "You are trying to tell me this armchair can fly?" to which the child replied, "It can fly, Papa." His father, after contradicting Andre's childhood ideas, tenderly, gently said, "Who knows, perhaps it is you who is right." André believed our modern civilization had become materialistic, impressed only by wealth, machines, efficient bureaucracy; but he always "just believed" there was something beyond, and not because he was told so. He just believed it, sensed it.

When he finally got to Bermuda again in 1947, he got himself a modest house with a studio to work on his ceramic art, often inspired by the Bermuda landscape and the sea creatures in Bermuda's coastal waters. He also gave Russian lessons for which, he reported to friends in a letter, he was well remunerated.

While he lived modestly, he resumed his old friendships from his winter in Bermuda in 1933, many of them interesting people. One was a British-born artist of Belgian and Scottish descent, Émile Antoine Verpilleux, who had initially gone to Bermuda in 1932 for a holiday but enjoyed painting there so stayed 18 months. It was at this time that André first met him and Verpilleux had been among the friends who gathered to see André off when the *Dal* left for New York. Verpilleux returned to Bermuda in 1936, married an American, and stayed. He did volunteer to serve his country's war effort, and was assigned to work for British Intelligence. Upon his return to Bermuda, he returned to art and by 1946 was exhibiting frequently at a new art gallery in Hamilton. Verpilleux was one of Bermuda's most successful artists and co-founder of The Society of Artists of Bermuda. When André returned he and Verpilleux resumed their friendship.

Just two years later, in 1949, André and Verpilleux collaborated on a set of large historical dioramas depicting scenes of early Bermuda history. Purchased by the Bermuda Trade Development Board for display at Fort St. Catherine, the dioramas were an important work of art at a time of increasing appreciation of Bermuda artists. Bermuda's high humidity unfortunately damaged the dioramas but as noted by Patrick Paley, a staff member of Fort St. Catherine, "They were the favourite of Fort Saint Catherine's displays. They are now a big miss for all those who remember them. The dioramas were a great visual aid for understanding Bermuda's historic events. I believe there were nine of them."

It was during this time that he met a sophisticated, multilingual woman who was holidaying in Bermuda. Almost the same age, they shared many interests, among them, books, music, history—including the history of Bermuda. And both, in their own way and for different reasons, were in search of a new beginning.

Kate—Cornelia Geertruida Sophie Cohen Stuart—was born in Scheveningen, Holland on September 12, 1903, to Sophie Christine Alide Engelbregt and Arnold J. Cohen Stuart. She was the eldest of four children.

Kate's mother was born on March 9, 1873, in The Hague, to a family descended from French Huguenots who fled persecution in France during the most violent period under Louis XIV's Edict of Fontainebleau in 1685. Highly skilled silk

Kate and André in Paris

weavers, the family changed its French name, de Beaux Arts, to Boesaart, to adapt to their new homeland. A hand written story of the family left by Kate's sister, Victoria, indicates that little else is known about the family's early years in Holland though Sophie's father, Pierre Engelbregt, who had been raised in an orphanage until the age of 16, had risen to a position of some prominence as head of the Dutch Department of Finance indicating on the one hand that Pierre was a gifted and disciplined young man, but also that he entered a society where merit was valued and humble origins were not a liability.

Her father, Arnold J. Cohen Stuart, was an engineer and a lawyer, as well as an excellent mathematician and a distinguished economist. A progressive man, he wrote his doctoral thesis at the University of Amsterdam on "Progressive Taxation" in which he argued that justice is achieved when each taxpayer incurs an equal sacrifice, to mean that the sacrifice of each taxpayer should be in the same ratio to the total satisfaction which each taxpayer derives from his income. This permits a higher rate of progression in taxes than John Stuart Mill's own version of the principle, and was widely accepted in the Edwardian era[1]. On the strength of this thesis he earned several nominations to professorships at the University of Amsterdam. He practiced law and served on the County Council. He joined Royal Dutch Petroleum in 1900 and after a short period of service in Indonesia, Arnold was appointed a director of Royal Dutch in 1906 and assigned to London.

His progressive views extended to the education of women and he recognized early on his eldest daughter's academic gifts, especially in mathematics. Kate's early education was at a day school in Hampstead, but at 14 she was sent away to a boarding school. With her excellent academic standing and her father's encouragement she sat for exams to Cambridge, still unusual for a woman in those days. She intended to study mathematics. However, after her father's sudden death in 1921, her mother put an end to all her plans and ambitions. Instead, she was to stay home and help her mother as her aide, travelling companion and hostess. By her sister's account, Kate felt "done out of higher learning."

Her mother later sent her to a "finishing school" in Paris where she became fluent in French and studied the piano, and then went to live with a family in Sweden, and learned Swedish. She returned to England when she was 21. She did volunteer work helping women in the East End of London, and in 1928 married a scholar of architecture a generation older, one of her parents' friends. This discrepancy in marital age was not unusual after the First World War when such a high proportion of young men of her age had been killed. The couple settled in a house in farming country thirty miles north of London, where they raised four children and Kate volunteered with the Women's Institute.

Although England was never invaded, in 1940 it certainly seemed about to happen, and the government encouraged families with small children to leave. Her own brother, Arnold, had been killed early in the war when shot down while piloting a Bristol Blenheim over northern France. Her mother responded to this loss by donating a large sum to the RAF to support the production of Spitfire planes.

Leaving England was a difficult decision but Kate managed, arranging to take her children to Canada on her own. She passed the war years in Toronto. With children at school, she worked for the Netherlands Relief Fund of the Red Cross under the patronage and participation of Crown Princess Juliana who had also found refuge with her children in Canada.

It was a different life, giving Kate a new independence. She organized and led family summer holidays horseback riding in the Rockies, the final three summers at the Brewster ranch at

Kananaskis, Alberta. She found the informal way of life and the outdoor experience of western Canada very appealing.

Among André's papers in the archive at the Piłsudski Institute in New York was a somewhat shortened version of his Atlantic voyage on the *Dal*. Written in Bermuda in 1953, in perfect idiomatic English, one can assume Kate and André worked on it together, although, prior to its discovery, the family had not known of its existence. One can also assume that at this time in Bermuda Kate told André about Kananaskis. Was this when an idea for a new start, for two people, was born? A new venture in ranching by two 50-year old Europeans does not seem entirely sensible, but one can't enter the hearts and minds of others.

Kate was divorced in 1949, she and André were married in 1951, and in 1953 they entered Canada on July 25, their "port of entry" in Lacolle, Quebec; their British passports were duly endorsed as "Landed immigrants.

Their first ranch, "Arrowhead," not far from Kevisville, Alberta, was a large enterprise covering 3,000 acres of land, half of it under cultivation for grain and hay as well as grazing land for cattle

Arrowhead ranch

Cattle at Arrowhead

Kate and André at Arrowhead Ranch

and sheep, and stands of timber that provided both income and met all the needs of the ranch. There was plentiful water with two trout streams, year-round springs and numerous water holes. The climate was not too harsh, summer days always cool at night, and winters mostly sunny. Arrowhead was hard work, even with a ranch manager and ranch hands.

It was during this time that André was made an Honorary Colonel of the Alberta Dragoons. He and Kate received invitations to official functions, enjoyed visits from various Canadian officers and André kept in touch with some friends from Poland and with his comrades from the Inter-Allied Commandos in England. They also found time for art, working together on a multi-media work depicting the Four Horsemen of the Apocalypse.

After two decades of ranch management, in the 1970s they sold Arrowhead and "retired" to Coleman, in the mountains of southern Alberta, where they established Bohomolec Ranch just above the town with a dramatic view of Crowsnest

Mountain. A very different enterprise this time, the much smaller ranch was intended as a retreat but it evolved into stables breeding quarter horses. Prize horses. By this time, André was in his 70s and suffering from the first signs of Parkinson's disease. Health issues aside, it was an unusual location for two cosmopolitan people their age to have chosen. It was beautiful but decidedly isolated and mono-cultural.

Perhaps because of his isolation in Coleman, he seemed to turn more towards a contemplative life. His thoughts took a philosophical turn. It was here he wrote another short book of a philosophical nature, *Mon ami l'Archange "G."* In it, he discussed his view of God and the Universe as One, and chose as a literary device a conversation between a Lieutenant and "G," presumably Gabriel. Different levels of knowledge and consciousness are attained through spiritual growth, in opposition to materialism. Many religions, Buddhism, Judaism, Christianity and Islam share an essential vision in common of a *Univers Divin = Deité Universelle*, though they differ greatly in details. It was also here that he sculpted a large, rather severe head of a man.

Not that he remained constantly contemplating metaphysics. It was at this time that he received a letter from some young sailors in Poland who solicited his help in erecting a monument to the *Dal* and to him, and about the same time he learned that his beloved yacht was to be remodelled in Chicago.

To the young sailors, he politely replied that it is neither fitting nor elegant to erect and pay for a monument to oneself. However, he was clearly touched by their proposal and found an acceptable alternative. His solution was to dedicate a chapel at St. Mary's Basilica in Gdansk to Mary, "Our Lady of the Seas," patroness of all sailors. He had very specific ideas about this statue. She had to have a young face in the style of a Botticelli, because, as he explained, the world belongs to the young. He engaged two artists, ceramicist Stanisław Konieczny and Janina Karczewska.[2] The plaque identifies the donor simply as "The sailor, Andrzej Bohomolec."

At about the same time, André had discovered that the *Dal*, after being displayed at the Chicago World's Fair was moved to Jackson Park and then to the Museum of Science and Industry where it remained for 26 years. In 1959 *Dal* was moved to the Polish Museum of Chicago and ten years later it somehow was turned over to private hands. Why this was done remains unclear, but André did learn that the new "owner" was about sail it back to Poland. Enraged, André resolved to get the *Dal* back to Poland, safely, to the Marine Museum in Gdansk where it remains to this day. But he had to wage a two-year court battle to get it back. The *Dal*'s transport to Poland was arranged by the Polish government.

Canada's official declaration of multiculturalism, in 1971, inspired André to write to the Prime Minister, Pierre Elliott Trudeau, to commend him for this. His letter to Trudeau is missing but the box containing his personal papers included

Dal at the Museum of Science and Industry in Chicago

an acknowledgement from the Prime Minister's Office, some draft notes and some correspondence with his friend, Piotr (Peter) Czartoryski, on the subject. He admired Canada immensely, writing that it must never become a "melting pot" that requires all newcomers to be thrown into the pot and boiled until all aspects of their identity have been boiled away, finally emerging with a generic identity. In his letters to Czartoryski, he writes about the value of languages and that all immigrants enrich Canada by bringing these to their new country. He valued languages for the cultural enrichment they provide, but also for their usefulness in trade and diplomacy. A diversity of religions living closely side-by-side would also add to cultural enrichment and fosters ecumenism. All of this, of course, with a common citizenship sharing common civic values.

André at the Crowsnest ranch

He loved Canada, a country between two seas, founded on two nations, where people of many ethnicities and religions live side by side, all of them citizens of a united Canada. People who know the history of his beloved Polish-Lithuanian Commonwealth would recognize that he was fondly describing the kind of country he idealized and fought for, but lost.

Endnotes

1. https://prabook.com/web/arnold_jacob.cohen_stuart/953323 accessed Feb. 13, 2019

2. https://www.google.ca/search?q=kaplica+ludzi+morza&tbm=isch&source=iu&ictx=1&fir=wW-ZpTtgf4YdLM%253A%252CpDD3Nxnhlzzc2M%252C_&usg=AI4_-kTbO9fkL3gUU1TkUGMmoh505aBTmA&sa=X&ved=2ahUKEwjawYOT7dfeAhUNnlkKHdjNC9YQ9QEwA3oECAYQBg#imgdii=nk170Tburr58zM:&imgrc=4FxOlsNqZ4gLKM: Accessed Feb. 13, 2019

Sculpture at the Crowsnest ranch

1

Preparation

My addiction to travel and adventure took hold when I was a child, largely under the influence of my parents who sent me out of the country every year during the winter. When they later sent me away to school, that too was abroad. With the army, I found myself in France—so when circumstances after the war made travel beyond my homeland impossible, I found this fate hard to accept. I found the solution to my restlessness by going to work on merchant vessels and that gave me an opportunity to visit South America, the African coast, India, Indochina. Along the way, I realized that if one wants to really see something while travelling, there is no need to be an unthinking, plodding tourist, the target for all sorts of travel agencies that seduce you to join their flocks to be herded along well-beaten paths. One must liberate oneself from the tedium of the routine, to avoid at all costs ordinary modes of travel such as trains or large steamships from which one sees nothing, experiences nothing.

Travelling further inland from the ports was largely out of my reach due to the cost of train fares and accommodation, and passport restrictions were another factor. Little by little, an idea took form in my mind for a long journey on a yacht. This idea first occurred to me earlier, when I visited the National Maritime Centre in Gdynia where I met my present collaborators, Jan Witkowski and Jerzy Świechowski.

I remember it as if it happened yesterday: I arrive at the Centre late in the evening, take my things to the tent, place them on the bunk assigned to me, and step out again into the yard. It is a beautiful night, though not moonlit; some 20 metres away from shore I see ghostly silhouettes of several yachts, their masts swaying lazily, hear the gentle waves lapping against the pier. The starry summer night is magical—what more do I need to inspire dreams and arouse my wanderlust?

A nearby voice interrupted my thoughts: "Very glad that you have arrived." I turn and in the darkness I make out the shadowy form of a young man, perhaps twenty-something. "Witkowski," he introduces himself, "I am the supply officer here." Then he adds, "Know what we should do? Let's take two of the yachts, and shake up some young people who have gathered here. There's a nice breeze, lovely night don't you think? I'll take *Kaper*." And to entice me further, "You take *Gryf*."

A crew I've never met, a yacht I've not even seen in daylight, unfamiliar with local conditions—but I don't have a chance to protest, Witkowski is already on his way. I hear his voice from afar, calling, swearing, and ordering what they are to bring on this nighttime expedition, though finding everything at once was not easy.

A quarter hour later, I am on a rowboat with my crew heading to board *Gryf*. Nearby, I can already hear all the voices and the general knocking about indicative of preparations for embarking out to sea.

"The sails are in the cabin, the lights in the forepeak and the lines... I checked them out this morning myself... watch out! There's a strong wind and waves smashing against the pier," he shouts.

In semi-darkness faintly lit by a swaying lantern, we raise our sails, and unmoor... On the other side of the pier a brisk breeze tips us to the side and we can hear the water pounding against the side of the boat as we gather speed and the harbor lights dim in the distance.

The next night we sail to the Hel peninsula, then later to Orlówo, then to Jastarnia. A year later, we sail to Sweden and Copenhagen with General Zaruski[1]. And then we launch a plan for a grand project.

It was during this time that I meet Świechowski, the youngest but also the coolest and most important of us three. He wasn't yet the great sailor he is today having not yet sailed on yachts but he was already known as an excellent navigator and had worked on steamboats. We are to learn later that he well deserved this reputation when we were caught in the hurricane that blew us far off course; we survived the hurricane and he expertly guided us safely to Bermuda having only a general map of the Atlantic Ocean and, as a chronometer, a pocket watch that he scrupulously wound every day, even on those days when we were shut up in a cabin half-filled with water, constantly tossed against one another and desperately waiting for the end. But that's still to come.

I started looking around for some kind of yacht for a longer expedition. My plans were beginning to take shape and I engaged the interest of some people who could influence the outcome. We were to be given a large yacht to sail off to the farthest corners of the earth... meanwhile things turned out quite differently. I was quite unexpectedly called to join a cavalry regiment in the *Kresy*, the eastern borderlands.

Well, that certainly made our plans go up in smoke since I was sent off to a distant part of Poland where I had difficulty communicating with Warsaw, let alone with Gdynia. The people who promised us a yacht abandoned us but, abandoned or not, our spirit was not broken. I corresponded regularly and frequently with Witkowski and Świechowski. Our plans once again crystallized but in a somewhat different form. Since we can't get help from anyone, we will simply have to arrange everything ourselves.

I had, at the time, a couple of thousand *złoty* at my disposal; that's the amount I was expecting to soon get my hands on. It was time to look around for a yacht. Easier said than done.

We examined every possibility we could think of. We needed a strong yacht, in the style of the cutters used by fishermen on their long voyages. Unfortunately, we found nothing of that sort. We set off on long and tedious searches outside Poland; Witkowski even went to Norway, but to no avail. Everywhere there was always, "yes, but..."—the boat was too old, the price too steep, the repair too costly.

But finally, I felt that I found the object of my dreams—a beautiful 2-masted ketch with a steel hull. About 13 metres long, it had a motor, luxuriously appointed interior, and a reasonable price. I was sure I finally found what I wanted.

Thinking I already had what I needed, it was time to approach my superiors for time off and permission to leave the country; they immediately gave me their enthusiastic approval. General Dreszer, as president of the Polish Maritime and Colonial League went so far as to transfer me to the National Institute for Physical Education in Warsaw and arrange the required permission to leave the country on May 1, 1933. We were then already in the month of April.

I will never forget my conversation with the general, which took place in his private home on *Aleje Ujazdowskie* [Ujazdowska Avenue]. It was only at this point that I admitted to him that our plan was to cross the Atlantic. (Until now, we only hinted about a long voyage perhaps around the coast of Europe.)

The general's expression became very serious; I nearly died of fear that he would now deny his support.

"My dear sir," said the general after a moment of serious reflection, "only two weeks ago Captain Skarżynski was here and told me he wanted to do exactly the same thing. I can only do what I did then—and that is to wish you all a successful voyage."

We are almost there! Just a few more minor details and we're ready to go. Or so I thought. But just a few days later some news hit me like a bolt of lightening.

Everything one buys abroad has a duty imposed upon it, yachts included. Unless, of course, there's a special trade agreement with the country of origin. Well, because of its particular country of origin, my yacht would not be exempt but be evaluated and taxed accordingly.

This is an awful blow. What am I to do? I was totally counting on this yacht. A tragic situation—on the one hand I had made a commitment to this voyage and on the other, I have no possible way of being able to do it.

Well, too late to back out. We are going, even if we have to sail in a bathtub, we are going.

I immediately inform my friends about this unexpected development and at the same time ask them to once again check out the possibilities in Gdansk or Gdynia. I don't wait for a reply but send them a telegram that I am on my way.

They are both waiting for me on the platform at the station in Gdansk, but it is only on the tramcar on the way to Westerplatte that they tell me they have found "something." But is it suitable? Well, a little small, a little tight for three, but if we have to find something right away then it will have to do.

Indeed, my friends did not try to mislead me. One glance and my blood turns cold at the thought of crossing the Atlantic in this. It is almost the same as the little sailboats anchored at the dock.

But it is affordable so what is there to discuss? A little bargaining, and I find myself the owner of this ship!

If I were to venture again on another long voyage perhaps I would choose an even smaller boat, but a boat designed like *Dal*? Not very likely.

Despite being well constructed with a strong hull, this little yacht is designed for coastal waters and has certain limitations for longer excursions into the high seas. For example, the hull is not designed as a "cutter," the front of the keel is situated under the mast which means in stormy conditions, tossed on a short but high wave, the entire front part of the boat, being flat, would stamp on the wave.

The skylight is leaky, the doors more like framed shutters than watertight doors. The rudder is too big, the deck—despite our reinforcements—is too fragile, the cockpit too big, the keel too light, and the inadequate sails difficult to raise and even more difficult to lower because they frequently get jammed at the top of the mast.

Whenever we broach far on one side, water leaks through the floor and into the cupboards along the side soaking all our supplies, clothing, and tools.

I know that anyone reading this would be inclined to be critical of us but those are people who would never venture beyond the Baltic anyway. They are like the crowd at a boxing match, munching on candies, shouting advice to the boxers and, later, confidently offering their "professional" opinions about the fight.

Most of our yacht's shortcomings were obvious to us before I bought it, the rest we got acquainted with before our departure from Le Havre. We tried to correct its faults, as much as we could.

But give up our voyage because of this? Never!

Our preparations began at a feverish pitch. We sailed the yacht to Gdynia where we put the boat high and dry in the fishery shipyard. While I rushed off to Warsaw to attend to the final details, my friends took care of the boat. First of all, the underwater part of the boat had to be covered with a sheet of copper that, among other things, protects the hull from seaweed and mussels sticking to the boat in southern waters. Considering that we would have to deal with great oceanic storms, several hoops around the hull were required to strengthen the construction. We added several steel lines to hold the mast. The original builder left space for an engine but we put in its place a large barrel for fresh water plus two more that were already installed beside the rear bunks that, along with a dozen or so containers of various sizes by the time we sailed weighed about half a ton.

Witkowski and Świechowski started making the storm sails. In the cabin we added leeboards to the bunks to prevent flying out when the boat rocked. And my kind and considerate friends even lengthened my 1.85-metre bunk to accommodate my 1.98-metre height. To do this they cut into the sail locker through which I could stretch my legs when I went to bed. As it turned out this space quickly got stuffed with all manner of things but I couldn't claim any right to complain.

Finally we painted and varnished the entire boat and lowered it into the water.

We spent a month in preparation and felt the pressure of time but there was always something more to be done. The storm sails were not finished, we couldn't get a sea anchor. But to finish it all always meant "just a few more days" delaying our departure.

Time pressed on.

Because of this we set off on the Baltic not entirely prepared for a voyage to America.

Sailing early we risked heavy storms on the Baltic and the North Sea. On the other hand, delaying our departure we risked the "cyclone season"[2] in the western stretch of the Atlantic. And under no circumstances did we want to risk that.

As fate would have it, we were not able to avoid either but for the time being we were optimistic about the future. Our determination to get to America before the hurricane season called for draconian measures. Enough! We leave in three days. We'll get the rest done along the way, in Copenhagen or Le Havre.

We filled the water tanks and sailed to the Officers' Yacht Club to load: preserves, dried fruit, vegetables, condensed milk, tea, coffee, sugar, bread, clothing, books, maps, navigation instruments, guns... Thanks to Mrs. Witkowska who generously offered her help, everything was in place, the bunks were made up.

There was an amusing incident during this time. While we were anchored next to the pier a lovely young woman accompanied by an elegant dandy with a foolish grin approached us. “Excuse me, gentlemen, where are you going?”

Startled, I replied, “Copenhagen.”

“Ach,” she said with a grimace. “and I thought perhaps Hel. Then you could have taken us along.”

She turned on her heel and walked away offended. But she may have had a point. Why sail across the Atlantic when you can calmly sail the waters around the Hel peninsula?

Endnote

1. General Mariusz Zaruski (1867-1941), was a pioneer in teaching Polish Scouts sailing, skiing and alpine climbing. Co-founder of the Yacht Club of Poland, he established the Sea and River League and organized Youth Marine Education on the Baltic Sea. Arrested by the Soviets, he died in prison in 1941.

2. A “cyclone” is a hurricane but in the Indian ocean.

2

Gdynia–Copenhagen

We raise our sails on June 5. Several groups of friends gather to see us off though they don't really know where we are going since we planned to keep our destination a secret until the last moment.

We hope to get to Chicago, because of the World's Fair and because there are some 600,000 Poles living there. How could we talk about such an ambitious undertaking; we don't even know one another well and all we have at our disposal is our boat. Rather than risk being ridiculed if we failed, we decide not to disclose our plans. This way, if by the time we got to Le Havre we felt we aren't up to crossing the Atlantic, we can give up.

The weather is fantastic, perfect for the start of our adventure.

We sail one more time along our familiar coastline, bidding farewell to our home port, around the Hel peninsula, and then out to the open sea.

Unfortunately, every coin has two sides. Not long after getting around Hel, at dusk, we are suddenly hit by steep seas cascading over the deck. Our yacht, which has not been out to sea in a long time, springs a leak at its waterline and water is pouring in. We start bailing. After a few minutes, our pump fails. We have to lift the floorboards in the cabin and use buckets to bail the water out of the bilges of the boat, a job that, given the strong swaying of the yacht and the narrow working space, is heavy work. This is when we find out why our pump failed. When we loaded our supplies, in order to use every possible space we stuffed boxes of preserves down in the bilge. When it got flooded, the water dissolved the glue on the labels and these little papers got stuck in our pump and jammed it. The long-term consequence of this is that with the labels gone we have no way of knowing what is in the cans. When we want vegetables, we open a can of lard, when we want lard we get jam.

By dawn, the sea is calm, the weather splendid. It is a perfect day for putting up an additional, large sail, which then briskly takes us out to sea.

We devote the next day to rearranging the cabin with great precision. *Dal* does not give us much room to manoeuvre. Because our loading was a rush job, and Mrs. Witkowska had helped, we have forgotten where things were put. It must be said that it is only in the second half of our voyage that we finally organize things more or less properly. At first, we stowed the most important supplies carefully and safely; as a result every time we need them it takes a long time to sort through less important items on top of them to get what we want. And, sometimes we find ourselves sorting through items a hundred times a day until we realize that we will never need them.

We are constantly buffeted by the waves and this affects our little yacht so much more than it would a large ship that some of our supplies seem to have acquired a life of their own. Some things calmly sway a little back and forth while others are constantly knocking about or worse, falling out of their designated place.

The next day we are on the wide-open Baltic where we have our first and most delightful encounter at sea. Witkowski is at the helm and we are about to begin a game of chess when we notice in the distance, on the horizon, the white sails of a rather large sailboat. After an hour of observation, we no longer have any doubts: judging by the size and type of sails we are convinced this is the *Dar Pomorza* [a Tall Ship for training scouts] and although it takes us off course, we nevertheless turn to approach her and greet our compatriots.

The sea is almost completely calm. Hoisting our large additional sail, which is made of the heaviest and densest canvas, we sail toward the great three-masted Tall Ship seemingly motionless on the smooth surface of the sea.

After exchanging a greeting with our semaphore flags, we are close enough to exchange a few words. What? Who? Where to?

But soon our voices don't carry so they reply in semaphore: "We wish you a good voyage."

And little by little, *Dar Pomorza* sails away, growing smaller and smaller in the distance, until she disappears from view.

Later that evening we spot the first lights on the island of Bornholm. I am at the helm, my friends are asleep, so I can daydream as much as I like. I feel disappointed that I won't see the island by daylight. By the time the sun rises again, we'll be too far away. I recalled my first time on a yacht, when I sailed to Bornholm with General Zaruski.

The cozy old port in Rønne, the warm-hearted people on that island, at first glance mostly fishermen, fair blonde women, and cyclists... and then my daydreams come to an end with a great bang on the bow of the boat. My heart stops, and before I can do anything I spot, in the darkness, a large beam floating beside me.

At this point, my thoughts take another path. I realize that the success of our venture does not depend entirely on us. Sometimes an unexpected detail, some blind chance, and everything could end a disaster.

My thoughts are then interrupted by Witkowski who comes up to replace me at the helm. I slip down to the cabin, lie down on my bunk, and fall into the sleep of the dead.

The next day we are approaching Copenhagen. We move slowly following a designated way for all ships around the many islands and rocks and drop anchor close to the Royal Yacht Club early in the morning.

We spend the morning putting the boat in order, and cleaning ourselves up. It is not until around ten that we are dressed and shaved, brass, paint and lacquer all glistening in the sun.

We can finally step out on land.

Copenhagen is truly a Mecca for sailors. In some respects, everything is done to accommodate a yachtsman. First of all, you enter and leave the port without any formalities. Neither customs nor police have ever been known to board a yacht. As for equipment, there is no other place that offers a greater choice. You walk into a yachting supply store and you want to buy everything. But if we were to buy everything we would have liked, or thought we needed, there would be no place for it in our poor little yacht.

We decide to put off any purchasing and instead I hasten to attend to officially announcing our arrival by first presenting our visiting cards to the Royal Yacht Club.

The club member on duty who shows me around the premises—and the club is splendid, especially the interior—proudly points to the pennants of affiliated clubs. Among them I spot the pennant of the Polish Yacht Club.

From there I go to register our presence at our consulate where they very obligingly stamp our ship's log testifying to our presence in Copenhagen.

Witkowski and I then attend to purchasing supplies. In view of the unfortunate incident with our pump, we buy another to have as a spare, plus 15 kilos rope, some oil for our compass lamp and some semaphore flags. A pressure stove on gimbals to enable us to prepare meals even on rough seas was our final, very important purchase.

As soon as we return to our boat we change out of our whites back into our denims and get to work. We have to install our new stove, change the ropes and pour oil into our compass lamp... but instead of giving us oil, they had sold us purified kerosene.

So what do we do? The stores are closed but we can't go out to sea without a compass light. Cursing the earth we stand on I stuff my bare feet into some old tennis shoes, put on a cap and rush off to town with an empty beer bottle under my arm.

It is dark so I walk ahead not knowing where I am going. My knowledge of the Danish language is primitive at best. How do you say oil in Danish? Probably something like oil, or oeuil, or ohl... I think it best to approach a policeman and pointing to my bottle try to explain to him what I want. The kindly officer of the law nods, smiling indulgently, and points: straight, then turn right, then left... "Restaurant something –or-other." I am dazzled by this man's good sense. Of course, at this hour the only place to get some oil is at a restaurant. I didn't realize that a certain brand of beer had a similar name. As it turns out, the directions take me to a very small square just behind the customs offices. In a central position is a tiny house, its function is impossible for me to determine. Between this house and the restaurant there stands a huge table. The main doors of the restaurant open into the square. Ten rather heavy people are gathered around this table. At the head of the table sits the chef, with a chubby, ruddy face his head topped regally with a gigantic white hat, along with several very handsome and elegant gentlemen, one with a little beard similar to that of Napoleon III and another, clean-shaven, busily sketching something on the table cloth. But for the moment, only one of them is of interest to me, and that is the chef. I approach him directly and once again repeat my need of oil for our cooker. The conversation stops and they all look at me as though at a lunatic and a moment later all of them burst out gleefully as I stand totally disoriented while the chef claps his hands on his considerable belly and shouts out an order toward the restaurant and the others pull me down to sit between Napoleon III and some beauty. The beauty, cheeks heavily rouged and dripping in jewels, has me drink directly from her glass but, startled, I tilt back and half of her drink spills into my lap. As this is happening, someone brings me a stein of enormous proportions filled with some very dark beer. The misunderstanding clears up but, unfortunately, too late... the fun has begun.

The chef, I learn, spent several years in Nice where he was apprenticed and acquired his culinary skills. The lady next to me once performed in Gdansk. Napoleon knows several dirty expressions in Russian, which he acts out, making the ladies squeal with delight while they implore him to stop. A couple then arrive in a large and luxurious vehicle and join us. Without interrupting the conversation, people are getting up and going either into the little house, or into the restaurant.

Very late at night I finally return to our boat carrying a bottle of well-earned oil...

3

Copenhagen–Ostend

We set off on the next leg of our voyage the next day at noon. It is fairly difficult to manoeuvre out to sea because the yacht basin in Copenhagen is rather on the small side and crowded with yachts, the wind is blowing against us and, worst of all, a group of seasoned yachtsmen from many different countries are observing our every move with great interest. A German crew that arrived the night before didn't take their eyes off us. It is essential that we make no mistakes and carry out our manoeuvres expertly.

We pull ourselves on the anchor line, hoist the sails in an instant, and changing tack three times in the narrow basin we successfully make our exit. Our national flag, a parting gift from our club, hovers joyously above us while next to the mast our international code flags signal our farewell while our French flag indicates our destination. After a traditional sailing around the port we move onward into the sea. Not long after we get past the fortified islands that protect the port we spot the familiar silhouette of the *Dar Pomorza* that due to calmer winds used a motor instead of sails and bypassed Copenhagen.

The wind is dying down. We have set the jib, so the light breezes move us gently forward. Then the sea turns completely calm, our boat rocks gently, our sails flutter limply. We take advantage of the calm to get everything into order. With our new pump, we thoroughly pump the water out from the bilge; we tidy up the interior; we scrub the deck; we look over our inventory in the forepeak because some exotic odours are emanating from that source. With great sorrow we discover a sack of potatoes that we brought from Gdynia has rotted. We spread them out on the deck, discard the bad ones, and spread the rest in the sun to dry. In our rush to pack, we didn't check the type and it turned out to be the worst sort.

With our work done, we bathe and turn our attention to preparing dinner on our new stove. The old one we toss overboard, together with the rotten potatoes.

In the evening, as we sail past the historical Kronborg Castle with its great fortified walls and concealed cannons pointing threateningly at us, we dine for the first time at sea: soup, potatoes, meat, fresh vegetables, compote from our collection of preserves, and the essential cup of tea.

It is here on these fortified walls that Hamlet encountered the ghost. The Polish baroque chronicler, Jan Pasek, wrote about these wonders in his diaries.

We are now 500 km from Gdynia.

It is time to assign watches that would remain in place for the rest of the voyage. From midnight to 4 am, Świechowski takes the helm; from 4 am to 8 am, I take over and then Witkowski takes over from me until noon, when Świechowski once again resumes his watch at the helm and thus we continue the 4-hour cycle. Of course in a storm or some other urgent matter we are all on deck as needed.

That night, under Witkowski's watch—before midnight—an incident occurs that could well have ended in disaster but then, as many times later, we played a defensive hand.

We were both sound asleep when Witkowski notices ahead of him the shadowy form of a fishing vessel, driven by a motor but without any lights. Sailboats always have the right of way when encountering a motor-driven vessel. Witkowski is using our manual light so that the fishing boat should see she is headed towards a sailing vessel.

But the fisherman, no doubt assuming he would encounter no one in this area is not at the tiller and his boat is advancing blindly.

The situation is critical. Not a moment to lose. Witkowski quickly steers hard to the side to get out of the way. But the fisherman, aroused out of a sound asleep, accidentally steers to port instead of starboard. Only a quick turn in the opposite direction at the last moment saves us from a catastrophe. Witkowski's colourful epithets and cursing, and the noise of the fishing boat's motor sounding as though right on top of us, we both leap up and out on deck... but the danger has passed. With philosophical calm, we return to our bunks.

We'll have something to talk about tomorrow, if weather permits a conversation.

When I emerge for my watch at the helm, the weather is magnificent. I like my watches, especially the early morning one when no one is on deck and I am on my own. In some respects it is not the best watch because the sleep before dawn is deep and very satisfying, but at the same time, this shift—4 to 8—has a romance about it: I am alone for both the sunrise and the sunset. Alone in my thoughts, it is a time of deep pleasure, a pleasure known only to hunters, mountain climbers and all who love to commune with nature. I gaze with wonder at the beauty of the Kattegat Sea.

A typical northern view: sharp outlines of rocks covered only with moss and pine trees with the sweet warm soft multi-colored rays of the rising sun shining on them.

But in all this I am forgetting to explain why we are on the Kategatt and not on the Kiel Canal that links the Baltic with the North Sea. It's simply because sailboats can't use the canal; it requires a motor, which we don't have. A boat could also be towed by a steamboat, which would be quite expensive. Consequently, we chose to sail around Denmark, a circuitous route and rather dangerous according to all sailors because of frequent storms, great rocks, shoals and many currents and crosscurrents.

One afternoon while changing the setting of the sails I step on the skylight and shatter the glass. This causes great consternation. We have extra glass... but nothing to cut it with. What to do? Turn off into some port and lose heaven knows how much time? We can't do something so foolish but neither can we leave it as is. We spend several hours attempting a variety of creative methods, all of them resulting in a great deal of broken glass. We finally succeed in cutting it to fit by soaking twine in naphtha, lighting it, and pressing it on the glass. Witkowski then installs it, replacing the shattered one.

In the evening we sail past the Danish island of Anholt. Because there are large underwater rocks in this area we use a probe to check the depth. Four to six metres, quite safe for us since *Dal* has a draft of 1.35 m.

At two in the morning I am awakened by my friends, neither of them was sleeping. Our situation has been unsettling since midnight—total stillness and fog—and we know there are shoals nearby, but where? Most likely strong currents too, but where? Our probes confirm our suspicions, showing depths of 7m, then 6, 5, 4, 3.5! Drop the anchor! By the time we have it ready, by the time we throw it, by the time it catches on the bottom, we'll be grounded on a sandbar. Not a moment to lose. We grab two big oars that we brought with us and we row with all our strength. We move at a snail's pace—after all we are rowing four tons—but finally the current no longer pulls us to the sandbar. We are counting on a breeze and finally we get one just as the sun is rising, and the fog lifts.

Every encounter at sea is very interesting and our next one a few hours later, when a warship—a small gunboat—belonging to the navy of His Majesty, the King of Denmark comes in view, our interest knows no bounds. As we pass one another we salute them with great solemnity lowering our national flag. The Danish captain, seated in a comfortable armchair at the stern of the ship waves to us amiably while a sailor, standing next to him, lowers and then raises again the Danish ensign. Etiquette is fully respected.

Before long we have another encounter, no less interesting than the previous one—I don't tell fish stories so you can believe me. At a distance of a mile or so we see an enormous flock of birds, a veritable cloud of them. Through our binoculars we determine, without a doubt, that they are ducks. We estimate roughly 2 or 3 thousand of them. None of us had ever seen such a large flock of birds. Where are they from? Local, or migrating? We are not knowledgeable enough to say.

That afternoon we sail past Skagen, Denmark's northernmost town. The weather is changing, the barometer is falling, the

On the North Sea

wind is rising, but nevertheless we continue not just with our mainsail but with two jibs! The biggest change is in the sea itself, the North Sea, which is actually an extension of the ocean. The seas are 1.5 m high but are short and violent. Our yacht can't easily sail on them; they are too short to sail smoothly from one wave to another, and too long to be able to climb over the waves. So she struggles, her bow at times forcing herself through, at other times collapsing with a deafening groan onto the next wave. From inside the cabin, this is a particularly frightening sound to the uninitiated. It feels as though someone is using a battering ram against the hull and will soon smash it. Our chances don't look good considering we packed no lifeboats for lack of space. On top of that, in the event we take in water, our half-ton iron keel will pull us into the abyss. We'll be nothing but air bubbles.

The waves are now washing over the deck—first time since we left Gdynia—and now we know that our cabin coachroof leaks. It seeps through the ceiling with great drops falling on the cots, and the occupants of the cots. Everything that is not secured tosses in every direction. Cooking is, of course, out of the question. This will be a long-term problem.

Cold, soaked, and hungry, Świechowski and I wait below, in the dark, until it's our turn to get into our oilskins to replace Witkowski at the helm.

After a few hours we turn south. On this course we get gentler seas from the quarter and the worst is past. Optimism returns to our hearts.

Actually, it would have been more sensible to keep closer to the coastline so in a storm we could shelter in the closest safe harbour, but time presses on us so we decide to continue straight as an arrow in a southwest direction figuring that in just a few days we'll be near the coast of France or England in the English Channel.

Next day, the 12th of June, something curious occurs. Despite fresh breezes in the rear and gentle waves, our yacht essentially is at a standstill, won't budge an inch. From our maps we know that the depth in this area is considerable so there can't be any shallows. We bend down overboard to have a look at the problem. We are trapped in some errant nets.

There's nothing for it. We drop sail, remove our shoes and using paddles and diving, we manage to liberate our boat.

In the evening we see signs of a storm, but with the characteristics of a land storm: darkness on the horizon, thunder and lightening. In fact, we are not far from land. We avoid one such storm by quickly changing course only to be hit directly by another. At first some bursts of strong winds, squalls. We were expecting that so are prepared with just one jib and the lower part of the mainsail rolled up—professionally speaking, we gave it two reefs. Heavy rain, thunder, lightning above us, then clear skies, good wind. We have to set sails again, once again hoisting the second jib.

Squall

Everything is fine except that we didn't get into our oilskins and we are soaked to the skin and on top of that, we broke one thermos. One less portion of hot tea at night.

That day, we covered 200 km. On a normal good day we do 180 km.

We are now about 1,100 km from Gdynia

During my next watch, the morning of June 13, once again the sea is completely still, not a breeze. A swell rocks us forcing the sails and all the rigging, from one side to another, which is harder on them and damages them more than under normal wind. To prevent more damage we lower the sails and start repairs—we change some of the ropes, sew up some seams on the sails. This takes about three hours and just as we finish the wind starts up again, as though on command. We hoist our sails and resume our voyage.

That day we covered only 100 km. Next day, before sunrise, we see masses of lights sparkling on the horizon. It looks as though we are approaching a city but we are hundreds of kilometers from land. Maybe this is a large fishing fleet. To avoid getting entangled in nets again we deviate away from them.

As daylight dawns we start seeing a steamship from time to time—a good sign. We are on course. We face a few navigational challenges. In order to estimate our distance from shore using the sun and stars to guide us—because we are using astronomy as our navigation guide—we brought with us various navigation books, charts and a sextant. Large steamships have very precise chronometers, which are essential. Unfortunately, since such a chronometer costs about 2,000 złoty [at the pre-1939 value] we had to make do with two old pocket watches that I had inherited from my father.

How could we have foreseen that although we had meticulously adjusted these watches under certain atmospheric conditions they would perform marvels of their own choosing?

At no time did they stay synchronized and in time we realized that one of them wouldn't keep time at all. Consequently, neither was of any use to us.

In the end, our navigator, Świechowski, takes Sun altitude at noon to determine our latitude, but longitude we have to estimate by dead reckoning (we don't have a log to measure speed).

Using this method we determine that our position on June 14 is latitude 54°17', longitude 5°5' East. Thermometer registers 14°, at night falls to 10°. We won't be taking off our sweaters any time soon. At night, the helmsman puts on an extra jacket and wraps his legs in a blanket.

In the afternoon we have another encounter with large logs probably washed off the deck of a ship carrying wood to be made into support beams for mines in England. We pull out a few and mount them on deck. One never knows what lies ahead and what challenges we have yet to overcome. Beams like this could prove to be useful: to reinforce our mast or any other beams if necessary, and in the event of a catastrophe, to have something to hold on to.

We are now about 1,400 km from Gdynia.

Next day we see a Tall Ship on our port side. This is our last glimpse of anything as fog rapidly envelops us and the winds die down. Once again we toss back and forth on a swell. Ships in the distance blast their foghorns, in accordance with international regulations (COLREGS) to alert others of their presence and prevent collisions. We follow suit, blowing our horn every minute or so.

The fog keeps thickening. The wind dies down, we are becalmed. I am at the helm, but despite the small size of our yacht and even though it is daytime, I can barely make out Witkowski crawling around the rigging in the bow. It seems he is in a bind. Sirens and horns are now heard with greater frequency so I assume we are not far from the English coast. Some ship is not far from us, struggling at half-speed. It seems so close, almost upon us. A large ship like that would crush us, without even noticing.

We breathe a sigh of relief as it passes us on the starboard side.

On the morning of June 16 the temperature continues to drop, there's a southwest wind but it doesn't disperse the fog. We're sailing close-hauled against a strong wind.

Among the foghorns we hear from many ships we notice a different intermittent sound. Checking our charts we realize this is the signal from the light vessel marking the Smith's Knoll shoal. So we must be close to the English coast. We are heading into the English Channel.

In the afternoon, the wind intensifies, the barometer drops at the rate of a millimeter per hour—there's no doubt: we will soon encounter something unpleasant. We have a couple of hours to prepare for battle. We begin by preparing and sitting down to a grand hot dinner. Who knows when we will be able to eat again. After that, we begin to reinforce the skylight, inside and out, fore and aft lids are tightly closed and the tarpaulin fastened. We leave only the jib, the mainsail is reefed down. That is half of what we had before though in fact, even this is too much. Our sails are not made of heavy canvas, not really storm sails. Unfortunately, our storm sails are not finished, nor is our sea anchor.

At 4 pm we get a short, sharp wave, the barometer keeps falling, the wind gets stronger, our little yacht struggles.

At nine o'clock that evening the storm is with us for good. In the darkness of night we could make out giant walls of water; they regularly fall upon us with a terrifying noise. Our little yacht revives and continues to fight for our lives. She falls into an abyss, then climbs over the ridge of the next one. Each time we get the feeling that this time she won't make it, she won't get over the ridge.

At midnight things are at their worst. We can't see a thing; the barometer has fallen to 746. The wind is merciless, tossing immense crested waves over the deck. *Dal* tosses on its side but after a moment, she straightens up and rises to the surface again. Manning the tiller is hard work. The helmsman has to be secured with rope lest he get washed overboard.

Heavy rain, ice cold sheets of seawater and foam splash relentlessly against our faces. Visibility is nil. All around us, and not far away, are great ships that, if they ram into us will crush us mercilessly without even noticing. And if they did happen to see us the most they could do is offer up a prayer for the dead because who could possibly find a survivor in this stormy hell.

It's impossible to see a man washed into the sea even in the light of day, so in the middle of the night...?

The bilges below the floor are full of water, leaking into the cabin soaking the bunks and seeping into lockers.

Exhausted and cold, Witkowski and I succumb to seasickness, and not long after, Świechowski joins us, without letting go of the helm.

Suddenly we hear a crash—the jib is broken, cries out Świechowski from the helm. It didn't withstand the force of the wind; we are without a sail in front. We have only one sail left, if a gale takes that one we will be just a toy bobbing on the waves. A dismal prospect.

At dawn, Witkowski is at the helm. The situation has gotten worse. To head the bow into the wave we must stay on course following the continental coast but surely the coast can't be far away by now. When we see it, we'll have to change course, an impossible feat in this hellish weather. There is a solution—rely on our sea anchor.

A sea anchor is a large canvas funnel on a long rope. Its function is to keep the bow of a yacht going into a wave. At least this is the theory that so far we have not tested. How will *Dal* respond? We decide we have to try this, but, since we don't have a sea anchor, we have to improvise. We break a small table and, using its legs and the storage bag for the sails, we are soon able to launch it.

Daylight. Cloudy. We are soaked to the skin, teeth chattering almost violently, we have no strength left. Silently we share the last remains of a bottle of rum.

The seam on our sail has split—yells Witkowski. Under our breaths we add—please don't let it split completely!

Our helmsman calls out: Yellow water! Yellow water indicates that we are in shallow water. So, not far from land.

We are prepared for the worst.

Then, quite unexpectedly, the wind gradually dies down. We see tugs and pilot boats with yellow or black funnels all around us. These are either Dutch or Belgian so we must be in a rocky estuary somewhere near the border between these two countries. We determine our position close to the light vessel, Wandelaar.

The sea is still stormy. From time to time we see the funnels of ships closest to us above the waves. One of them, a Belgian, wants to rush to our rescue. Her captain, standing on the bridge, is shouting through a megaphone but the wind drowns out his voice. We respond—thank you, but we'll manage. He then tries to frighten us, indicating with a stern expression that the storm is coming back, he is pointing to the sky... but we stand firm. We know the price we'll pay if we accept help: we can barely afford a third-class ticket to return to Poland! For about a half an hour the ship follows us but eventually the captain turns his ship around with a wave of his hand, probably adding: Go ahead and drown!

After brief deliberations we decide to sail to Ostend since the wind is once again not cooperative and besides, we don't have any dry thread left on board. Not that this is such a simple thing to do. As a rule, when a ship sails into port, the crew has a plan of the port and the surrounding area indicating every buoy, every rock, all the shallows, and then takes a pilot on deck when entering the harbour. We, on the other hand, have only a map of the coast and in a stormy sea it is not easy to orient ourselves from a low deck like ours.

Well, we have no choice. Head for Ostend! After all, we are in fairly shallow waters.

A couple of hours later we can see Ostend through our binoculars as clearly as the palm of our hands. The monumental buildings lining the boulevard along the coast are impressive—and we can see the beach, with masses of small cabins for dressing, some of them on wheels to easily move them to and from shore as the tide is flooding or ebbing.

We are at low water, the supports of the pier are above water and completely surrounded by seaweed. There is no way we can find the passageway used by the great ships. The usual way is for one of us to move fore and watch the water from the bow; if necessary, the helmsman can save the situation with a quick, sharp turn.

The sun is high, so we can more or less see what's in the water. We observe the passage of a ship and follow in the same direction.

A lot of vacationers are on the mole, all of them waving to us, and we to them. We suspect, we know, why they are greeting us so warmly: we look like we are survivors of a battle at sea! Our sails are torn, our rigging is hanging in disarray, we are dirty, unshaven since Copenhagen, wearing sweaters and oilskins. Our story is obvious at a glance.

Beyond the mole the water is smooth as glass, there are storm warnings on the signal tower. We feel comfortable and secure, as though sitting by a fireplace while a storm rages beyond the windows.

As we pass the many ships that have taken shelter from the storm we salute them with our flag, they return the salute. We turn around a few times, let go the anchor, lower our sails and, we are in the marina of the Belgian Yacht Club.

We immediately get to work, each tackling a task. There is no time for resting. Świechowski and I take everything out and lay it on deck, despite continuing, intermittent rain showers. Nothing will dry out right now but it will overnight. We are hungry as wolves; Witkowski is rustling up some food, getting splattered with hot grease and cursing the whole time, making fried potatoes. But he carries on regardless.

Dal looks a fright! We have blankets, sheets and towels flapping in the breeze, sweaters and shirts playfully fluttering their sleeves. The entire deck is covered with mattresses and pillows. I scrape the rust off the rifle—a gift from my brother-in-law. It looks like an ancient artefact retrieved during an excavation. Meanwhile, Świechowski, with a serious expression on his face is pouring the water out of his camera, humming a recent popular tune, off key.

Foggy and windy, with an occasional light drizzle, and yet there is such great joy in our hearts we are almost giddy, as we ponder...

It's a wonderful world!

4

Ostend–Le Havre

The storm severely damaged our yacht. On deck: the jib was shredded, the mainsail damaged; in the cabin: broken dishes and everything soaked. Our pharmacy, so carefully organized by my mother, was a major casualty, as was our little library—bathing in the sea was not good for them.

By noon, after more or less tidying up, after bathing and shaving, Świechowski and I disembark, our yacht documents in hand. There were still occasional squalls and rain. We enter the Royal Yacht Club, which seems empty until we encounter a chap in the hangar, a chubby, paint-splattered club member busily painting his little boat.

"And what were you gentlemen up to?" he asked. "It's 24 hours since you came in from the sea. We were observing you yesterday. I know! I know!"—he cut in before I could reply—"dealing with the damage from the storm. Still, after ten days at sea a man wants to put his feet on land. But where's the third chap? You're a crew of three, aren't you?"

When told that "the third chap" was in no great rush to get on land he shook his head but said nothing more. I got the impression that he would not be seeking out our company any time soon.

As for me, I must admit it is with great joy that I stroll along the boulevard by the shore, despite the occasional raindrops on my face. What a delight it is to be able to straighten up without banging my head on the ceiling of the cabin, as happens to me several times every day; to be able to wave my outstretched arms without hitting the rigging. Well, not an unmitigated delight... Świechowski is by my side, glancing askance at my show of exuberant joy.

The 20th century tourist approaches travelling as though he were mixing a cocktail, except that instead of pouring alcohol and tossing a few ice cubes into a shaker, he throws himself and a couple of valises into a cabin on an ocean liner, shuts himself in, gives himself a good shake, then pours himself carefully into a colorless, generic "Grand Hotel" with little interest in the country where this hotel happens to be. After a time he returns, his suitcases now covered with multi-colored stickers as evidence of the countries he visited and tells everyone: I have been to India, Australia... True, he did visit these countries we can't deny that, but for heaven's sake, does he know anything more about these countries than he could have learned just by browsing through a Baedeker?

If you haven't ever cast anchor at a foreign port with your own hands, have not refreshed your horse at a roadside well or at least changed the tires on your bicycle or other vehicle on some road a thousand miles away from Poland, then you haven't earned the right to call yourself a traveller. And you don't really know the pleasures of "travel," in the real meaning of the word.

Indeed, you can't even know the euphoria of the mountain climber who reaches the highest peak in the Tatras.

Once we registered with the harbour master, bought our provisions and sent off a few postcards (one can't after all, forget one's family, regiment and club!) we start wandering around town. We spend considerable time in the Gothic cathedral with its crypt and magnificent stained glass windows. Next, the fishermen's wharf where we admire the beautiful, massive cutters bedecked with nets, and every few steps we encounter fishermen from another era in their waterproof gear, shod in wooden clogs, earrings in pierced ears, smoking their short pipes and earnestly discussing the weather and the price of fish, in Flemish.

It being a holiday, the streets of the lower town districts are packed with people. A bazaar, with stall after stall on both sides of the narrow streets leading to the market square, its owners and attendants dressed in costume—a veritable masquerade: cowboys, Chinese coolies, soldiers of both sexes. One stall, decorated in a presumably Louis XV style, is selling some sort of exceptional shoe polish, and next to that, the owner—seemingly of Indian origin—extolls the quality and flavour of his candy. Everyone is noisy, everyone is in good humour. I don't know exactly when this happened but I suddenly noticed that Świechowski had been captured by wolves, sea-wolves that is, and together they present a lovely tableau around a long table stacked high with feminine clothing.

"Buy this one!" they say, knowing that one sailor cannot refuse a request from another, "your lady will be pleased." Displaying one garment after another, some of which modesty does not allow me describe. After much bargaining and the purchase of two scarves, we are finally liberated.

A bit farther a little crowd has gathered. To attract customers to his stand with a variety of clothing, the owner and his two salesgirls, holding hands, perform a folk dance, much to the delight of their audience.

Must be sure to suggest this to the Jabłkowski brothers who own a large department store, *Bracia Jabłkowski,* on Bracka Street in Warsaw. Another day, the weather unchanged. We think our barometer had water damage, it is not moving. In town, the reading is 5 mm higher. There is no meteorological station here; reports from Brussels are posted a day late. The port continues to show storm signals.

But enough of that! We can't just be stuck in this port. After a brief discussion we decide to leave the next day, regardless of the weather. Today, for the last time, the club bosun ferries us from our boat to land; this time Witkowski accompanies me. Port authorities, a stein of beer in a pub, final purchases, one last promenade in the tourist area along the beach, which is fantastic and wonderfully empty (must be out of season), and finally the bosun takes us back to *Dal.* I hastily express a few words of thanks for the hospitality asking that he extend them to the Commodore of the Royal Yacht Club, and then we are once again a company of three.

On June 20 in the afternoon, after a lavish dinner, we sail away in high spirits, taking advantage of the ebbing tide to take us out to sea. We have made sure we have everything we need in the event of another storm. The damage inflicted by the last one has been repaired, the deck is clear, what's left has been secured against waves that have no respect for other people's property; wouldn't want to let them get anything because of our carelessness.

The cabin was also secured with attention to every detail, the chain, ropes and anchor; things that belong above are stored in the forepeak. The fastenings on the bow and on the skylight are tightly shut and covered with tarps. The mainsail has been reduced in size, meaning its surface area is now smaller; if need be, we'll reduce it even more.

Away from the pier the weather is not as bad as indicated by the barometer and the constant storm signals. (Maybe the signals are a permanent fixture.)

About five o'clock in the afternoon the wind dies down almost completely. We take advantage of this to cook our supper using our fresh ingredients but our dining pleasure is compromised by several brief but strong squalls, coming one after another, which heel the yacht sharply on the beams. We just finished with the squalls and our supper when no sooner had we dealt with the squalls then once again we face the arrival of storms of the land variety with thunder and lightning.

Next day, Świechowski calls me at four in the morning to come up on deck. I can see from below that he is dressed in his oilskins. Bad sign! The yacht is rocking mercilessly and I hear the constant noise of waves washing over the deck and the wind howling in the rigging. So once again the weather is not cooperating.

I emerge on deck, the mainsail is reduced to the limit. Briefly and quickly we change watches, check our compass course, make a few minor observations about our situation and, wrapping myself in a blanket, I log the events of the previous night.

I am at the helm—difficult, despite the bright and beautiful day. I am not dealing with waves 5m high, as we did during the storm, but they are still 3-4m, and they are short, so the yacht struggles. As we sail past the East Goodwin Light Vessel (which marks the treacherous East Goodwin sands in the English Channel), the wind intensifies. Then a crack, like a shotgun. The jib clew line snapped. With nothing holding it the sail blows in all directions like a giant white flag; another minute and it will be shredded but after a few minutes of fighting with the canvas we get it fastened again. All is well.

Around four in the afternoon we are sailing past Calais and across the Channel we can clearly see the English coast. We are now in the Channel, the wind dies down a bit, the waves are smaller. Seems the elements have conceded that battle to us and won't bother us further. No such luck. In the evening we encounter another enemy, this time less frightening on the surface but nevertheless dangerous, underwater currents caused by flooding and ebbing. The Channel is generously endowed with these. Twice each day the ocean falls into it, as if into a bathtub, only to fall back again, six hours later, where it had come from. In the narrows, the current is no worse than in the Vistula in the spring.

As if to convince ourselves of this, in the evening, just as we pass Cap Gris Nez, the sea is suddenly becalmed. We know exactly what we can expect. When, in a couple of hours the current reverses itself, it will take us God only knows where, back where we came from or, worse still, toss us against some rocks or sandbars. We grab the oars furiously rowing trying to get to some place near the French coast, some two miles away, where we could drop anchor. But despite our heroic efforts we are moving with the speed of a tortoise and time is against us. Nightfall. The lighthouse at Cap Gris Nez, its signal starting only minutes earlier, seems to be slowly but surely giving up on us. Our hearts sink; we know at that point that the ebb had begun. We put down the oars and set to work to get the anchor ready. We take the heaviest anchor we have, attach the chain, then to the chain we tie our strongest, thickest rope and fix it securely on the bow. Once everything is ready, we let go the anchor and with great anxiety await developments.

At first, nothing... nothing... the line drops straight down... too deep... A short time later the line strains forward. The anchor holds! The line drops again... the anchor lets go. The line is taut... it's dragging! I picture in my mind our anchor, its flukes desperately trying to catch bottom but the merciless current keeps dragging it, dragging it... farther and farther.

Suddenly Świechowski interrupts my thoughts. This time the anchor holds for good. It's true, the line is taut. The hull sharply ploughs the current and were it not for the lights at stationary Cap Gris on one side, and at the English towns of Folkestone and Dover on the other, we could have believed that some mysterious force is towing us at great speed into a dark, moonless and eternal night.

At dawn the current reverses, accompanied by a good little breeze. We have to take advantage of this current over the next six hours in order to sail as much distance as possible.

Now the next problem: no amount of effort gets the anchor up. Must have hooked on a rock, and we can't budge it. We go for the most radical solutions. Using ties and levers we pull up the rope and half of the chain... and that's all we can get. The rest of the chain drops straight down, meaning we are exactly above the anchor. We pull so hard that even the stem points down, while the stern rises. Taking advantage of this all three of us rush to the back of the yacht; our collective weight lowers the stern, the bow rises and—crack!— the chain snaps. We hoist sails.

In our log for June 22: at 5:00 pm Cap Gris Nez, we lose our anchor and 15m of chain!

The weather is splendid. We strip and warm ourselves in the sun after a sleepless night. At the helm, I dream sea dreams. With joyful anticipation I think of the moment when I take Witkowski's place peacefully stretching out under the mast. A few miles on port side is the rocky coast of France, bathed by the setting sun. A quarter mile or so away a couple of dolphins leap and play. The English coast has receded, except for some smoke rising probably from steamboats.

Shortly after noon, we encounter another ebbing current. The breeze is too weak to go against it. We expect another critical time on our anchor. It is our last difficult event of this part of our voyage.

Next day, June 23 approx. 6:00 pm, we arrive in Le Havre and anchor in the outer port. Around midnight, they open the lock; we enter the inner port. This manoeuver is not so easy. Most yachts, as we observed during our two-week stay, when entering and leaving the port, make use of their auxiliary motors or are towed.

Since we can do neither, we have to enter a port with which we are not familiar using our sails, in the dark, but by now we are an experienced team so we manoeuver smoothly, pass through several large gates opened for passage and moor for the night until the drawbridge opens in the morning.

It is quiet, warm and cosy in the old port, all around us, along the sides, several small ships await with us for the opening of the drawbridge so they can enter the main port. Very small old houses are tightly circled in a ring around us, their dark silhouettes against a starlit sky. There are old-fashioned gaslights casting a hazy light on the sidewalks and on the windows of several pubs. Through an open doorway one can see a classic old nickel-plated bar and hear the knock of billiard balls, the splash and clatter of dishes being washed. It reminds me of something... The same décor, the same atmosphere of a small bar in a port... Yes, I saw it at a Polish theatre not long before I left. *Marius.* [A drama by the French playwright, Marcel Pagnol.]

Interesting. As a rule, art portrays reality. Now it is the other way around, reality reminds me of art.

But really... do we always know what is truth and what is illusion?

5

Le Havre–Plymouth

We have now spent two days moored in the main port, moored not at the pier but alongside a powerful, modern pilot ship, *Senateur Brindeau*. This comes with some great advantages. We can make use of some benefits on the ship; put some of our things in a covered area of its deck; the ship also shelters us from the rising and ebbing tides which can be felt even here and we rise and fall in tandem with the ship so we don't have to keep adjusting our moorings or worrying that the hull will scrape against the shore. Oh, and most important: we are protected from curious passers-by and uninvited guests.

The captain of the pilot ship despite having the appearance of a typical, honest shopkeeper is an old sea wolf and he, along with the entire crew, treats us very kindly. He is very talkative and proud of his ship which he took command of only recently.

"Yes indeed, gentlemen, this is the most modern ship, cost a solid couple of million; we live like kings here. But nevertheless," he adds with a bit of melancholy in his voice, "it wasn't so bad in the old days when we were real sailors. In those days, a pilot owned his own good but modest sailboat, maybe no more than twice the size of yours. We used to sail out to meet ships no matter how great the storm... but let me tell you gentlemen something that will amuse you: I began my career on *Number Nineteen*! You must know her; it was Poland that bought her."

Do we know her?! Of course we know her, she was a solid and valiant boat in the yacht club in Gdynia. We sailed many a voyage in her.

The port where we are docked, despite its name, "le Bassin du Commerce," is used almost exclusively by yachts. In our spare time we wander around and, examining every yacht with a critical eye, we passionately admire or disdain, as the case may be. And every time we announce our opinion, before launching into great detail, our pronouncement begins with the words, "Now, if I had the money to... etc. etc."

Just as elderly ladies at a ball know who's who and with whom, so we too quickly know everything about our neighbours.

Right next to us is a huge motor launch, painted white from its keel to the very top of its funnels. It resembles more a commercial liner than a luxurious pleasure craft. Previously the property of an Egyptian or a Persian, it was then called the *Ismet Pasha*, but some workers are busily removing the brass lettering on the hull. The new owner, Mme C., who after obtaining a divorce, together with a fortune worth millions—from her husband, a famous maker of perfumes—ordered a change of name. It is henceforth to be known as *Qu'importe*. I would give anything to find out what provoked this choice of such an emotionally disturbed, desperate name; was it after a long but futile search for an appropriate name or simply the prosaic appearance of the ship itself.

Immediately next to Mme C., a long row of smaller yachts, some, among them, first rate. Then a yacht that had won a number of Atlantic regattas, then a steamboat belonging to one of the Rothschilds, a little smaller and more graceful than *Ismet Pasha*.

But the best and most beautiful of all is the majestic square-rigger, *Manou*. Every time I go past it as I attend to some errands I can't resist stopping for a while just to gaze at this fantastic yacht. It belongs to a Mr. Bourgeois, the owner of the transport company C.I.M. (Compagne Industrielle Maritime). What a coincidence that I happen to have a letter of introduction to the director of this company, Mr. A. Laffon, to whom I am related through the Paris branch of my family.

After walking past several basins and bascule bridges, there is a huge hangar at the end of the mole where I am received by the potentate of the transport company in his office. When the superintendent brings me to the office I am pleasantly surprised to see before me not, as I had expected, a severe old man but a cheerful thirty-something fellow, a passionate yachtsman who greets me like a brother. He was aware that we were on our way but had not worried about us having received word from a ship on the North Sea about the storm, and that we had made it.

Having the same outlook, the same tastes, and many mutual cousins and friends in Paris, it soon felt as though we had known one another forever.

"What can one do?" he admits a while later, "my misfortune is that I chose what is called 'a career'. The position comes with monumental responsibilities. Before you know it, a fellow is chained to his job. I would infinitely prefer to live a carefree life in the South Seas. I will do whatever I can for you. No need to thank me because, for my part, I will do so out of self-interest. Maybe some day I too will be able to break away from Le Havre. And then I would want that my fellow yachtsmen in distant ports would welcome me with open hearts."

And so he does. The next day a motorboat arrives from C.I.M. to tow us to the company dock; there, before we are even aware of it, an enormous crane lifts *Dal* like a feather and sets it on shore on a scaffolding already in place for it. For the crane, this is just a trifle since it could easily raise 400 tons. Our work takes off. Thanks to the help obtained from the Polish Consulate, notably Mr. Tyszko, who helped and facilitated every step of the way, we are able to enlist several unemployed men from the local Polish community. The entire boat is thoroughly sanded and a new coat of varnish is applied, the storm sails are finished, the deck is painted. We buy a new anchor to replace the one we lost at Cap Gris Nez, lots of rope, and loads of good stuff: shackles, screws, nails, metal-work and carpentry tools, needles, sail thread, etc. etc.

When everything is more or less finished I can no longer resist and take advantage of the opportunity to hop in a car and go to Paris for a few days.

Seven days later, when my sister drives me back to Le Havre, *Dal* had already been put back in the Commercial Port next to a pilot boat.

Time is pressing. We attend to our final tasks, I secure our provisions for the journey.

Since deliveries are made to the deck of the pilot ship next to us, I break up the crates and pass items to Witkowski using the ramp.

We realize only then how much we can pack into our little yacht, despite its slender proportions: 60 cans of beef, 120 tins of condensed milk, 150 cans of canned vegetables, tomato sauce, lard, butter, marmalade, compote, a 30-kilo crate of dry biscuits, 50 bottles of wine, 40 packages of macaroni, as well as rice, vegetables and fruit, cocoa, tea, coffee, tobacco for Witkowski, and kerosene for our stove. We still have a bag of sugar and several litres of purified alcohol. We buy three large barrels that enable us to increase our water supply to 150 litres.

But we are missing things in the navigational department. When we arrived in Le Havre we knew that one of our watches was kaput while the other was temperamental. To remedy that, we buy a small radio from the telegraph operator on the pilot ship, just an ordinary radio receiver of the type one would find in any home.

With the aid of this radio we are able to receive, twice daily, the radio signal from the Eiffel Tower giving us the exact time and thus regulate our temperamental watch. The problem of keeping time is now, more or less, settled.

We also buy maps of the channel, the Bay of Biscay, Madeira and the Canary Islands, a general map of the Atlantic Ocean, and even the Antilles, just in case we find ourselves in that area. As it turned out, we had no use for any maps except those of the Atlantic and the Channel.

As I already pointed out, when we left Poland, we had no definite idea what our goal would be because we hadn't tested ourselves yet as a crew, nor tested the yacht, so it was too difficult to know in advance just what we were capable of. We saw two possibilities: if we felt that we were not up to a battle with the ocean, we could just sail around Spain and then follow the coast the length of the Mediterranean.

As for the Atlantic crossing, the situation is this: we have no desire to sail north from the Channel because we have had enough of that on the North Sea; taking the most direct route to America would give us a lot of trouble because the prevailing winds would be constantly against us, so the shortest route would not be the fastest. We would have to sail south of the Spanish coast to get the Trade Winds that are most favourable for an Atlantic crossing.

But again, there are two variables.

For the easiest way, we would have to go south along the coast of Europe, and then Africa, crossing by Cape Verde and the narrowest stretch of the Atlantic. A more difficult but faster route, now rarely taken by sailing yachts to North America, requires making a great circle. This route goes across the open sea immediately upon leaving the channel. After brief consideration, we choose this one.

I don't want to bore the reader with details about the formalities in Le Havre, with the Port Authorities, Customs, or the Polish Consulate. However, I must say a few words about our dealings with the American Consulate.

In order to enter the United States, our yacht, like every other ship, had to obtain a special visa, a rather expensive proposition. To enter the United States, the Americans demand documents declaring that the port of departure is free of infectious diseases and, moreover, they demand to know whether any passengers or crew have among them persons who were not permitted into the US.

At the US consulate I was handed a form the size of a bed sheet on which you could list the name of the entire crew of a battleship. The picture that came to mind as I looked at our three names listed in a corner of this massive form was that of three sleepy passengers disembarked from a night train in an otherwise empty monumental train terminal of a major city.

Finally, equipped with a map of the ocean with meteorological markings for the month of July, and with our watch now properly regulated to keep accurate time, we are ready to set off.

First, our farewells. Several yachtsmen come to shake our hands, among them the nicest and most entertaining is the very young owner of one of the smaller yachts in the basin. This energetic little person looks very distinctive since, not having with her something sporting to wear, she is dressed in an afternoon dress with a little purse under her arm, and standing in a microscopic boat she calmly skims across the entire basin propelling her oar with the skill of an old sea fox.

Mr. and Mrs. Tyszko, from the Polish consulate, come on deck to drink a toast and wish us a good journey. Nor did the American Consul forget us. The last one to leave us is Mr. A. Laffon who arrives on a company motorboat, stays with us until we pass the lock and only then bids us farewell.

We sail away from this hospitable port under full sail on a beautiful day on July 8 at eleven o'clock.

Among our repairs and reorganization in Le Havre we carried out the following innovations: a radio to receive time signals, its box carefully placed at the foot of Świechowski's bunk, its batteries in a cupboard next to that, grounded on the brass trim of the hull and the antenna pulled up on the mast to receive the signals. In the cockpit at the helm, we had constructed a large crate for our carpentry and metalworking tools, liberating a lot of space in the cabin, while reducing the size of the cockpit considerably. We obtained a large square tarpaulin, a gift from the senior manager of the C.I.M. Company. This tarp was to enable us to shelter the exposed part of the cockpit against the waves during a storm.

The very next day the weather is clearly changing for the worse. Every little while we reef our sails. The wind keeps intensifying and turning so much that we have to constantly change tack. Short waves, howling like a pack of wolves, toss us in every direction. Once again the window above the cabin, the "skylight," starts leaking, the water dripping on our cots. Slowly everything gets thoroughly soaked. Cooking is out of the question.

We have a companion in our misfortune, a large yacht that zigzags under the force of the wind, just as we do.

By evening the storm is in full swing. Our two-week holiday on land has weakened our resistance. We are all sick. My stay in Paris, especially, did not strengthen me.

After a brief discussion we decide to head for the nearest English port, to Plymouth.

The situation is serious. The storm, plus nightfall. We have only one general map of the English Channel in which the area of Plymouth is the size of a dime. The map only tells us that to enter the port in Plymouth we must pass the lighthouse from the left.

As it turns out, even this information is false because there is a huge breakwater and massive waves blocking the view of the port. The breakwater is not marked on our map so you can imagine our horror when our prow is suddenly in these roiling waves. We make a sharp turn and continue alongside this barrier. That takes us to the end of the mole so a few minutes later we are sailing at a regatta tempo (strong winds, no waves) following the shore. We can see a few lights on shore and a few yachts rocking lazily nearby; this must be a marina. We let go our anchor and furl the sails.

A little while later we are drinking hot tea, our first warm food since we left Le Havre. Everything on the deck is in order, the anchor is holding fast, we can go to sleep.

When we wake up, we look around to see what there is to see by morning light.

About a hundred and some metres away there is a high, rocky cliff with a walk carved into it. In the centre, supported by posts extending out to the sea, a hideous, massive glazed building. A tin dome, some columns with tin ornamentation. Something like this could only have been created by extreme bad taste around the turn of the century. There is an identical building in Nice, from the same era. Some wit aptly dubbed it "created in the imagination of a delirious confectioner".

Then, a surprise. Anchored at the other end is the yacht we saw last evening. They must have waited on the open sea until dawn before coming here to shelter against the elements.

The wind has turned and is now blowing in from the sea. We are now dancing wildly at anchor, dinner preparation would require skills more acrobatic than culinary.

The weather here is identical to that in Ostend. Every few minutes the wind tosses a rain shower at our faces. An officer from the port authority stops by and without shutting off his motor asks us the name of our yacht and our port of origin. Some young people come by in a rowboat for a brief visit. At this time, none of us can say very much in English so our communication is mostly in sign language. They invite us to the yacht club. Unfortunately by the time we get there, in the evening, the clubhouse turns out to be in town rather than by the shore and in any case, it is closed.

We wander around the town. For some reason unknown to us all the restaurants and pubs are closed. Świechowski and I therefore have no choice but to go to some confectioner's shop and eat a couple of bland little cakes, watching the rain on the windowpanes. Some peddler tries aggressively to sell us some ladies' hosiery, by the dozen. Stockings... cakes... rain... It's a dismal world.

When the rain clears up a little we walk to the park. From its manicured lawns to its monuments and cannons, everything is faithful to tradition. That same old and tired tradition that insists that everything be measured by the yard with its 3 feet, the currency in pounds each worth 20 shillings each worth 13 pence, tennis is played in white trousers, and riding horseback requires red jackets, playing bridge and going to the cinema is forbidden on Sundays, and everything that is not English is, without question, second rate.

The next morning our boat is rocking so mercilessly that we decide to go to the fishermen's wharf where it will be safer. Unfortunately, all our efforts to break out an anchor come to naught. The bosun of the club explains to us that there is a thick chain on the sea floor here and that our anchor is likely fouled on it. The prospect of losing yet another anchor and 10 metres of rope does not make us happy. That would be another considerable expense.

At six, we postpone all decisions and sit down to eat. Just then we have the most foolish and at the same time most dangerous moment on our yacht.

While we three are sitting in the cabin below, a thought comes to me seemingly out of nowhere: seated closest to the hatch I stand up and look out. My hair is standing on end! Whether it is because we pulled in too much of the line from the anchor, or because the wind changed directions and unhooked the anchor from the chain on the sea floor, or it was the tide—enough: the fact is that the tide is propelling us towards the rocks which are only about 50 metres away. In the blink of an eye we are all on deck. Within moments, splash—and the second anchor is overboard. We are racing against time... and the second anchor doesn't hold. We are helpless against the inevitable, with no strength left. Use the oars to save ourselves? Can't be done; the current and wind are too strong. Raise the sails? Too late. The sails are furled; it would take a full five minutes and we are seconds away from crashing against the rocks, sure as two plus two makes four. I pictured in my mind a smile, and words at once sweetly ironic and preachy: "I told you so, you little specks."

But it seems that sometimes in life two and two does not make four, but rather five. That "five" appears to us out of nowhere, the way it does in some American films. At the point when all seems lost, the young girl is found, her fiancé has not been killed, and all is well that ends well. Our happy ending comes in the form of a passing motorboat.

Some shouting, some tossing of ropes... and half an hour later we are calmly being moored between fishing vessels on the deliciously filthy and calm water of an inner marina.

We are doubly lucky that we were not smashed on the rocks, nor were we thrown to the merciless law of the sea. The law of the sea demands that a ship has an obligation to rescue only the crew but to rescue the vessel as well could cost a considerable sum. Often enough, during the worst storms, tugboats will gather around the crippled ship like jackals.

Then the bargaining begins and only when the price is settled is the towline thrown.

Had our rescuers wanted to take advantage of the situation, they too could have demanded a couple of thousand before throwing us a line, which I would have had to either pay or lose our yacht.

Now that we can go ashore more easily, we go several times daily into town to check the meteorological communiqués so that we can get to the Bay of Biscay in the most favourable conditions since Biscay is famous for its storms.

My heart seems to be bothering me so I take advantage of this free time to get a check-up.

I find a young doctor—and judging by the photographs decorating his office an avid sportsman. He speaks French, listens to what we had already done and what we plan to do, checks my blood pressure, listens to my heart and lungs, writes out a prescription, and says: "You've survived up to now, you will survive the rest; you can't, after all, keep having bad weather."

So I am reassured about the state of my health, as are my two partners. Witkowski, who was constantly complaining about his teeth, already had them pulled wholesale in Le Havre—I'm not sure how many. Świechowski had been sick with a cold but is now feeling better.

While in Plymouth, we buy a big crate of oranges that, for lack of space, we simply place in the middle of the cabin, along with a crate of biscuits. Also some fresh food for the first days at sea, some preserves, one electric lamp, new batteries for the radio.

We are ready to sail.

July 15, four in the afternoon, our sails set, we sail out of the fishermen's wharf, the fishermen gathered on shore waving farewell.

We have left Europe. Will we make it to America?

6

The Atlantic Crossing

I am in deep trouble—that is I've been in trouble all along—but now, I have no idea how to get out of it. That is, how to chronicle our adventures from this point on. When it came to writing about our voyage along coastal waters, I somehow managed it, because our collective psychology had not changed from our days on land; but after leaving Plymouth, despite the majesty of the boundless ocean surrounding us, our constant confinement on our tiny shell was gradually changing us. During our coastal travels, what we found interesting could also be interesting to a reader. But things are different now. Our primary concerns now are the details of daily life but it is difficult to write page after page that Świechowski is drinking his tea without sugar, that I am diluting a tin of tomato sauce with water while the others think it should be used as is, that Witkowski insists on smoking in the cabin while Świechowski and I vigorously are opposed to that, the sugar got spilled, and so on and on.

So I have no choice but to ignore these tiresome details, these minor squabbles, these minutiae of our daily life and this way give our voyage a certain idyllic character that in reality was not exactly like that. I hope my readers, especially the young passionate ones, keep that in mind.

The weather at sea continues to be disagreeable. On July 17, the winds die down a little but the rolling sea is not yet ready to calm down, the waves slam the topsides from every direction.

While my friends are sleeping, I am alone at the helm deeply absorbed in my work when I suddenly hear a frightening noise, a snorting and a splash. I turn and freeze. Some 20 dolphins had barely missed us.

The dolphin is the least dangerous animal in the world—never even attacks swimmers let alone boats—but still, their classic cavalry charge complete with an element of surprise scared me to death. These entertaining creatures, some of them half the length of our boat, encircled us in no time, like a disciplined, well-trained squadron, pulling southeast.

By evening, the wind dies down, fog descends; the strong morning seas from all directions gradually weaken.

At eight o'clock we adjust our watch for the first time according to the signal from the Eiffel Tower. At this point, let me provide some details about our radio equipment, for the amateur radio operators out there. As I mentioned before, we came across it in Le Havre and decided on the spur of the moment to buy it. It's an old-fashioned battery operated radio with radio waves ranging 300 to 3000 metres. Every day, weather permitting, we attach a single lead 6-metre antenna to the mast, its other end attached at the helm. With this primitive arrangement we are still getting the signal a month later, some 7000 km away from Paris.

When sailors see herds of dolphins swimming together in one direction they usually say that they are going with the wind, and sure enough, early next morning during my watch there is a fairly strong southeast wind blowing. On the other hand, the sea has calmed down. So we unfurl a full sail and start moving at racing speed. In the afternoon we put our soaked belongings out to dry in the brilliant sun and devour a meal that could have easily fed ten men. It seems everything is in order, yet we can feel stormy weather in the air, the barometer falls a little, and we are in the stormiest bay in Europe. As I learned later, almost an entire fishing fleet went down in these waters just a few months later. We come upon just such a fleet later that afternoon.

Large, stocky two-masted sailing boats—ketches. On both sides they have long poles each with rows of rods. To prevent the sails from rotting, they dye the sails using some red or blue preparation; with these colourful sails they distinguish themselves from other ships.

One of them turns off course and sails so close to us we can smell the fresh fish. The owner, a pleasant, chubby Frenchman, points up to the sky and shouts a warning: "Weather!" We wave our thanks, but he quickly sails away.

His warning is not immediately borne out. However, in the afternoon, the seas get bigger and start flowing over our deck; we must quickly gather our things that were drying in the sun and get them into the cabin, below.

In the cabin, we begin to notice a strong smell of mustiness, of spilt kerosene, wet blankets. The oranges are rotting. I look over them, throw about twenty overboard. We have to eat the rest as soon as possible. We eat an unbelievable quantity. Witkowski sets the record, one day eating 18 of them. He is punished accordingly.

The last of our vegetables, half a sack of potatoes, the macaroni that we had stored in our clothes cupboard, all are unusable. Toss them overboard! My poor old alarm clock that has faithfully timed our naps and watches, suffered the same fate. He put up a good fight but ultimately rust wins the battle.

About the same time one morning, we notice an unfortunate detail. Our mast, which never pleased us very much because it tended to sway, is slightly bent. We have no desire to head for some Spanish port to replace it. We finally decide that we will straighten and strengthen it using steel wires—stays.

From that moment on, we change watches in the morning observing a certain ritual: when I hand over the helm to Witkowski each of us looks up and examines the mast scrupulously. "Well?" one of us then asks. "Like a candle!" is the routine response.

It should not be surprising that we show such concern for our mast—its loss would come with serious consequences. Back in Le Havre we received a letter informing us that one of our best yachts was lost precisely because of a broken mast. As the mast fell it holed the hull causing the metal keel, several tons of it, to drag the boat down into the abyss.

But in our case, even if our mast breaks and does not damage the hull, we'd still be in big trouble. How would we replace it, and how much farther should we go? There are no pine forests growing in the ocean.

At last we are out of the Bay of Biscay and out into the great ocean, about level with Spain's north coast. The weather is beautiful, with pleasant fresh breezes and we are at full sail. Our little yacht is speeding along as if on wings, we are all in excellent spirits. At sunset, we should take down some sail for safety. Unwilling to lose speed, we decide to leave it at full sail but to double our watch. We achieve our objective, not only because we don't lose speed but also because we gradually set an incredible pace.

Against the midnight blue, star-studded sky, our two large sails, the mainsail and the balloon sail boom out on both sides, in the darkness they seem like a huge bird. Running dead before the wind, the wind blowing more strongly, the boat has the appearance of a barge on the Vistula. It feels as though we are jumping from one wave to another while the sea throws back large bursts of foaming water on both sides of the hull.

Sailing this way makes great demands on the helmsman. The slightest deviation off course could tear the sails, damage the rigging and, God forbid, an accident with the mast.

In the morning, the early light reveals to us some weakness in the sails. There is a tear in one seam of the balloon sail. We lower it and change the rope... and we do this just in time. There's a change in the weather, it's getting stormy. We reduce our sails further.

We have emptied one barrel of water, and our bread is so mouldy that we throw it all overboard. From now on, dried biscuits will be the staple of our daily menu.

At first we don't really know how to deal with this because their texture and their golden-yellow colour reminds us of the tiles covering the stoves of long ago; but little by little we get used to them. At times, we even like them. Snacking on these dry biscuits, soaked in tea, wine or even seawater, helps pass the time when we are on watch, especially for Świechowski and me, since neither of us smokes.

Witkowski's smoking, however, presents a new problem. Not sure it is because the tobacco is damp, or because he isn't a skilful smoker, but he goes through half a package of matches whenever he's on watch. This situation takes an even more dramatic turn when we discover that we had been sold not kerosene for our stove but contaminated alcohol impossible to ignite with our lighters.

Świechowski and I very sneakily hide all the matches. To prevent our smoker from falling into utter despair we offer him a magnifying glass and that works at least during daytime shifts.

We are now ten days on the ocean, the temperature constantly rising. Instead of jackets, sweaters or blankets, ordinary shirts are adequate even at night. But as mentioned before, the cabin is stuffy, hot, damp and stinky. So we eagerly resort to a minimalist wardrobe as we rock gently back and forth on a swell and empty the cabin of everything including mattresses, books and even floorboards and give everything a thorough scouring, scrubbing, rinsing and drying, as required.

By the time our work is finished we are filthy, so we dive in and swim around *Dal* glancing frequently left and right on the lookout for the sharp fin of a shark slicing smoothly through the still water.

We cover only 52 nautical miles on this windless day but next day we make up for lost time sailing 128 miles.

The same day we have a playful encounter with a colony of huge turtles that is following us without any sign of fear or apprehension. The expression in their eyes as we sail only a few metres away from them reminded me of the gentle and rather blank look in the eyes of cows as they look at a passing train.

At this point we are more or less between Madeira and the Azores.

At last we sail into the NE trade winds, the steady winds that at this point on our planet blow constantly from Africa to America. To encounter a storm or to be becalmed is a rare thing in the trade winds; on the other hand 4-5-metre waves are very common. However, they are so long and gentle that one hardly feels them. The danger will come when we leave the trade winds and approach the American coast.

It is the beginning of August; the sun is merciless. For over a week we have dispensed completely with clothing except for big straw hats to protect our heads from the sun's rays. We

In the Trade Winds

Swimming in the sea

cover our bodies with oil and somehow get accustomed to these conditions. Within a short time we are tanned but with a most bizarre pattern, that is, only our backs, shoulders and knees. We haven't shaved in who knows how long, all of which makes us look much more like pirates than like sportsmen.

Little by little, our rigging and sails are getting worn. Every day something tears, bursts or just shows wear and tear. At times we need more rope, other times we have to repair torn sails, replace the wire rope. Essentially, we should have two sets of sails but we have only one and it was used by the previous owner for two years. On the whole, the sails are holding up pretty well, especially the biggest one—the mainsail—which only burst a seam a couple of times and which we immediately repaired lest the seam give way against the wind. The daily routine is now pretty much set. After my morning watch I have some tea and then attend to some bosun tasks. Also, at this time, we make entries into the ship's log, i.e. I make note of direction and the strength of the wind, our speed, our course, which sails are in use, the barometer reading, the weather, the sky, the sea, the temperature, and the more important events of the previous day.

I am also starting to keep notes about our journey. Then I go to sleep. In the afternoon I am usually awakened by the pleasant aroma of our dinner. Świechowski's watch gets the full force of the sun. At four o'clock we once again calculate our position. In the evening, after supper, we get the antenna up—always a tense moment and we hold our breath: will we still get the signal from the Eiffel Tower? For ten days now we have not received the morning signal. In the evening, Świechowski rests ahead of his next watch while Witkowski and I play a game of chess on my pocket chessboard. Later, when it gets dark, we remain on deck sipping tea and gazing at the stars, or the extraordinary pale blue phosphorescence of the sparkling bright and bubbly sea below.

When we leave the Bay of Biscay three seagulls that have been keeping us company the whole month join us again. They are a special kind of seagull with small, three-cornered heads and exceptionally long tails consisting of just a few very large flight feathers. They are the only living beings on this endless watery desert. One is black, another white, and the third—also white—has a sharply lobed tail. We consider them good friends. Sometimes, around noon, they fly away for a couple of hours but always return and circle mournfully above us. We feel they are watching over us.

There was a time when sailors believed these birds were the souls of sailors who had perished at sea in tragic circumstances.

We are now in the tropics and the sea gives us magnificent views—always the same, yet constantly changing. I'm not referring only to sunrise and sunsets with the magnificent daily spectacles of fiery crimson but also the entire sky in variegated and also constantly changing colours. Sitting several hours at the helm, you can observe many interesting phenomena. The sky is so intensely blue it hurts your eyes. There are only the smallest clouds above us but all around us are tiny, balloon-like clouds scattered along the horizon. Schools of flying fish are all around us like native sparrows. Every so often we hear the splash of large fish—tuna and dorado (dolphinfish) as I later learn. We are now frequently encountering a new variety of jellyfish that the British nicknamed "Portuguese Man o' War;" with their large, brightly coloured, oval protuberances cresting above the water they seem to be made of Venetian glass. Although I knew better, I could not rest in peace until I had fished one of these jellyfish out of the water. This was no easy undertaking. The movement of our boat creates a wave that tosses the jellyfish away from us, making it difficult to reach them. At last my long desired moment comes. A perfect specimen appears right beside me. I leap from the stern, grab him with my hand and throw him on the deck. If I were given the opportunity to do this again today I would prefer to reach for a wasps' nest with my bare hand than tangle with a jellyfish. For an hour I howl with pain; my arm is numb all the way to my shoulder. It's most interesting that this leaves no permanent symptoms, neither burning nor swelling. Taking revenge on my behalf, Witkowski murders the creature most cruelly, burning himself in the process. The worst thing, however, is that the entire left side of our boat and the ropes where the jellyfish had lain, are covered with her corrosive fluid, so much so that despite scouring, rinsing and drying in the tropical sun, for the next couple of days we get burnt every time we touch it.

This experience effectively cools my interest in zoology.

On August 8, there is a great celebration on the deck of the yacht, *Dal*, of the Officers' Yacht Club R. P. [R.P.—Republic of Poland]. We are at the mid-point of our crossing of the Atlantic, an achievement that calls for a bottle of wine. It is strange to realize that there is not a speck of land anywhere around us within thousands of kilometres.

Our radio signal gets weaker each day; finally on the 11th of August I have to record in the ship's log: the radio no longer receives the time signal... nor can we reach American time signals.

August 12. Great excitement. A steamship!!! Judging by its course, we decide it is South America–New York, flying an English flag. Passing us at a distance of a couple of miles, they send a friendly signal: "25°N49'W". We reply in kind, using the international code flags along with our national flag: "Please radio our position."

The next day another steamship, but this one is far on the horizon. For sure they didn't see us. Judging by their course we decide they are sailing England–Antilles.

Very strange clouds in the sky; we've never seen anything like it. An art exhibit of clouds. They are white, round, innocent shapes, the kind that saints and angels sit on in religious pictures. Some are long, feathery streaks; some woolly sheep; trade wind clouds on the horizon; and behind us, dark rainclouds. Thunder!

Next day, the exhibit is still on display above us but more dark clouds coming, from every direction. By the afternoon the clouds are so black that we don't see the setting sun. I want to insert here the difference between a storm on land with thunder and lightening, and a storm at sea—in Polish there are two different terms for these storms; on land they simply call it "nasty weather" but on sea, a gale, with winds that create very strong waves.

Well, I've heard from many seamen that there are no electric storms on the ocean because such a vast expanse of water levels out the clouds as they come. When we encountered a couple of storms on the North Sea, we tried to get closer to land. But here, in the middle of the Atlantic at its widest stretch?! Were it not for our later adventure, this storm would have been the worst part of our voyage.

It starts in the evening, around six. The sea is as flat as a mirror. The barometer is stable. Earlier, at noon, we try to take advantage of every light air but, unfortunately, they were very rare. Black clouds ominously covered the entire sky. The air was so permeated with electricity that we felt electric shocks in our fingers just from touching anything metal, or even wet instruments. Nervous tension beyond endurance. Just in case, we furl the extra headsail keeping only the mainsail and the foresail.

It's now completely dark because the storm, or I should use the plural—storms, because the phenomena are now across the entire sky. The clouds, though closely together, have not lost their distinctive appearance. We are struck by the power and majesty of nature. Thunderbolts strike all around us. They strike directly into the sea, or at other clouds, in extraordinary formations. Some spherical, then zigzagged, then radial. Then comes the heavy tropical downpour. Almost total darkness. Everything is totally still, except for the warm rainwater falling on us in streams. When the rain stops—and it stops as abruptly as it started—it is followed by a fierce wind that blows our yacht completely on her beams. The sails are almost touching the water... if they get submerged, we have no idea how we would straighten her out again. Lowering the sails in this squall, especially in this position, is impossible. We gradually pick up speed and soon, *Dal* straightens up. We are now sailing at great speed on a smooth ocean. Then more warm tropical rain. Taking advantage of momentary calm, almost blinded by rain, we lower our sails.

If we get more of such powerful blows, we are safe.

The next evening, more of the same, but not quite so strong. I forgot a few details from the previous night. Worried that the lightning could strike the mast, we grounded the steel lines with wire cast overboard. But at night we decided against this. Whether it was because of this, or something else, we note another phenomenon, which is more worrying. It is during Witkowski's watch. The surface of the sails is minimal, so we aren't concerned about strong winds. Świechowski and I are sitting in the cabin, actually lying down since we take a rest whenever we can. Who knows what the night will bring. Suddenly Witkowski is calling me on deck. Świechowski joins us. The top of the mast was shining brightly. None of us is an Edison or a Marconi. How then are we to evaluate whether this is dangerous? We watch as the light at the top of the mast gets brighter. Suddenly this blinding light is all around us. "This is the end," I think. But no, everything is fine... except that a little flash of light appears again at the top of the mast. These flashes of light keep appearing for the next couple of hours. The light would increase, increase until after a couple of minutes it is spent. We don't want to risk "grounding" the mast again since we don't know if attaching a wire in these circumstances could cause a shock. The charge proves to be inconsequential... but it sure tested our nerves.

At about four o'clock in the afternoon we have our next, and actually our last, close encounter with a steamship.

From the moment they spot us on the horizon they change course towards us. We begin a desperate search for our trousers and shirts which have been hidden somewhere in the depths of the cabin for several weeks now. The easiest to find is my shirt because we have been using it to dry our dishes.

We are overwhelmed by the sheer size of this ship, which effectively comes to a stop quite close to us. The side of it looks to us like a large steel wall decorated on top by a long row of heads looking down upon us.

It is the M.S. Frida Horn, from Hamburg, heading to the Antilles. The captain very kindly asks us if we need anything. We thank him graciously saying that our only request is that he make our whereabouts known. The ship's engines start up again and it moves away, all of us waving until we are out of sight. A half hour later, all we could see of the Frida Horn was a long trail of smoke on the horizon.

7
Hurricane

For readers who want to have a real sense of what we went through in the next few days after our meeting the *Frida Horn*, I suggest they read Joseph Conrad's *Typhoon*. I have to admit that I don't have the literary power to adequately describe our ordeal. Right from the start I would have to start using so many superlatives, words like "enormous," "powerful," "huge," that eventually my vocabulary would fail me.

My failure would disappoint my noble ancestor, Franciszek Bohomolec[4], and my professors who failed me in my Polish language exam at the military college must be turning in their graves that I have the audacity to put pen to paper.

I will content myself to write simply and factually, with quotations from our ship's log in which I scrupulously made daily entries. In order to be as accurate as possible, details, such as the force of the wind and the state of the ocean, etc., I only made these entries after all three of us discussed and agreed on them. The state of the sea we measured on a scale of 1 to 9. Nine was: The Furies unleashed. For wind force we used the international Beaufort scale of 1 to 12, twelve indicating that on land rooftops and cows would fly through the air with the greatest of ease.

August 18. Yesterday afternoon there was a wreath around the sun. We all know that this halo does not bode well but we all avoid saying so. Reluctantly one of us goes down into the cabin and raises the barometer up to the skylight. That's all. The clouds get darker and darker. Storm clouds gather from all sides. Squall winds measure 8. We are using storm sails, much smaller and of very heavy canvas. Waves rise to four metres. The barometer falls from 764 to 760. Once again we prepare for battle. Everything on deck is secured with ropes, as also everything in the cabin is either reinforced or shut away. Cooking is out of the question. We eat dry biscuits and drink wine.

August 19. The barometer fell to 756 overnight. There is no more doubt as to what awaits us. We are in the path of a hurricane. Monstrous waves, just the foam on the crest spans two metres. It's all too much for poor little *Dal*. Every minute or so a crest falls on the quarterdeck and floods the cockpit. The mast is seriously listing. In the afternoon our battle is against impossible odds leaving us only one option—use our sea anchor. It's an extremely difficult moment. We aren't sure just how the yacht will respond to the sea anchor. We set it, along with 90 metres of heavy anchor line by the cockpit. The other end of the rope we fix at the bow passing the line outside backstays and shrouds, wind it around a bollard and fasten it at the base of the mast.

When all is done we cast the sea anchor overboard and gradually let out more rope. Then quickly in less than the blink of an eye, we drop the sails.

Unfortunately, *Dal* is not a ketch so does not have a second small mast aft to which we could attach a small sail to help the sea anchor. We handle that by fastening a small sail to the boom, set on topping lifts just above the helm. The yacht wobbles momentarily but then stabilizes itself quite well. It's not ideal because we are not set exactly perpendicular to the waves but at a 45° angle. Still, we climb and fall off gigantic mountains of water, piercing the powerful crests on top.

For a moment we overcome our terror, and the incredible nervous tension we felt up to that point subsides. We think we have passed the worst. We are euphoric, experiencing a childlike delight. Witkowski, holding on to the deck with both hands, splashes overboard, while Świechowski performs other, somewhat more prosaic, acrobatic feats.

Nightfall, the conditions get worse, we figure the sea is at 9 on the scale, the barometer has fallen to 739, the gale winds are back, in the afternoon it blows North, now it blows NNW, the force of the wind is at 11. Although we head into the waves, the heel of the boat is still at 30°.

It's the second day without food, we gradually get depressed. The water continues to leak through the hatch and the skylight, flooding the floor. We have to pump it out constantly. The entrance to the cabin is blocked with our sails; Witkowski sleeps on them because his cot, near the hatch, is covered with water.

August 20. It's three o'clock. Barometer at minimum 736. Suddenly all is calm—we must be in the eye of the hurricane. Then, suddenly again, the gale winds abruptly turn 15-17 degrees, with a crack rip off the small sail at the helm as well as the canvas covering the hatch. Eight o'clock. The position of the yacht worsens, though by now there's no point in talking about our position because the wind is constantly changing direction (right now, west), its force is 12 on the Beaufort scale. Every change of direction of the wind also changes the direction of the waves; the opposing waves collide creating a watery hell.

We can't leave the cabin. It's dark, and will stay dark for the next three days, but through the hatch we get an occasional glimpse of the monstrous avalanches of water crashing against one another and then rising in huge columns into the air, the hurricane winds smash them and carry the scattered waves high into the black, angry sky above us.

We are heeling at 50° to 60°.

We are suddenly hit by the crest of a wave, tipping us more than 100°. Witkowski and I shoot up like bullets, taking mattresses, floor boards, rolled up sails and anything else that's not nailed down with us, hitting the ceiling above Świechowski's bunk, while foaming water floods in through the hatch.

But the ton and a half of ballast we have in the keel does what it is meant to do. After just seconds—though it seems an eternity—the boat straightens as quickly as it had collapsed on its side. We are sharply tossed back again.

The water in the cabin reaches our bunks. The same thought flashes through all three of us.

What is happening on deck? It seems to me that it took me just one second to crawl out from under boards and sails and get to the hatch but my friends were already there ahead of me lying flat on the deck hanging on by their fingernails and their teeth to everything one could possibly hold.

The foamy peaks on top of the waves extend three metres. The mast is gone, only a short, bare stump is left—the rest is in the sea, attached to the yacht by the rigging. With every toss the mast hits the hull like a battering ram and may well smash it at any moment.

We feel as though we are on some heaving, tossing sea monster rather than on a boat. The deck, sometimes lying steeply above me, other times steeply beneath me, and only when we emerge from a wave do I see my friends clutching the remains of the mast. The sea anchor is gone. Every wave turns the boat on her beams. It is so full of water, and sits so deeply in the sea that one can barely see its sides. If the cabin gets filled one more time as before, we are certainly finished. The pumps are somewhere under the water in the cabin. I reach in where the sails are... I remember there's a pail there.

Crawling on my knees along both sides of the hatch... tough work. We have long passed the limits of human endurance but we continue, and will keep on going. It's a matter of saving our skins.

Half an hour later, the water has been mostly bailed out, on deck we have come up with something, instead of a sea anchor we toss out a couple of beams, attached by a long rope. We are almost collapsing from fatigue. The yacht tips almost on its side with every wave.

In the cabin we have a feeling that the wind is slowing down. But that is only an illusion.

Usually, when closed in a cabin one can estimate the force of the wind by the sound of the steel wires against the mast, a singing sound at Beaufort 5 or under, over 5 a whizzing sound, rising and at around 10 more like the sound of a hand rubbing against certain kinds of silk fabric. At 11 or 12, sounds like the crunching of large sheets of packing papers.

But we don't hear any sounds, not because the weather has not changed but only because we have no mast. We are now positioned with port side against the seas. I am lying down, tied down actually, to prevent myself from suddenly being thrown against my companions; I can hear a sound from outside, a noise in the distance, getting closer and closer, transforming itself into a powerful roar and then we feel as if some monstrous power is lifting us, then a terrible blow, a bang and a shock. The yacht is wailing like a wounded animal. This is repeated every minute or so, tipping us on the side at 60° or so.

12 o'clock. Midnight. We are below, in the cabin, the hatch covered with canvas. Through a tiny opening we made just to get some air I can see, not the sky—but contrary to all logic—the foaming ocean above us.

Another hard blow, tipping us more than 90°, tossing us sharply, and once again the hatch is leaking. The superstructure is wrecked, the canvas on it is torn, the water is streaming in, through the ceiling, through cracks everywhere.

The situation is hopeless.

I have to note something commendable about us at this time: not one of us ever uttered a word of blame or complaint. Each of us in our own space prepared himself for our longest, final journey.

The barometer is rising quickly, 1/3 mm per hour. A small spark of hope enters our hearts. Since we survived the first half of this storm, perhaps we'll make it through the second?

But we have one more nightmarish night. We are constantly on the alert for the next wave and after every hit the same thought runs through our heads: Well, it wasn't this one... and then we wait for the next.

August 21. In the dim morning light we see the weather remains the same as yesterday; however daylight, though dim, cheers us up. The waves keep coming from all directions but the worst ones still hit on the port side. We get to work, pumping out water, fastening the oars, boat hooks, and beams on the left side of the quarterdeck to protect it from the force of the waves.

The barometer keeps rising rapidly, by noon up to 752 mm (the normal in this area is 764-5) and stops. The wind force now is only 8 B, though every so often we get a gust at a force of 11B.

We keep getting short bursts of rain but not enough to have a calming effect.

August 22. The wind force is falling, not gusting much any more. The seas no longer have foaming crests. The waves resemble great domed hills but still steep enough that they can flood the deck and tip the yacht to 50°.

We really don't look human any more. We haven't had anything in our mouths since the morning of the 18th, so four days without food, water or sleep. We have a great deal of work ahead of us. The hurricane caused the following damages: the 9-metre mast is reduced to a stump of 3.5 m, sticking out with nothing attached to it; the rigging that's left is lying on the deck. In fact the entire yacht is a scene of misery and despair.

Six o'clock. We begin work strengthening what remains of the mast. We have to first take it out completely from the step and attach the rigging to its summit. All the wires that secure the mast to the sides are at this point too long; we tie knots in them to shorten them.

The tiller, which is cracked at the base, must be reinforced as much as possible with wire. The work is exhausting, in part because we ourselves are exhausted, waves keep flooding us on deck, and *Dal* frequently heels 50°.

The sun appeared around eleven o'clock creating great wreaths of light all around us. We are intrigued ... what can it mean? Does it mean the hurricane has not entirely left us? Is it a warning about an encore that will be even more spectacular in its violence than the original show? But there's no point in speculation: we hoist and adjust the storm sail, get out the beams that substitute for a sea anchor, scurrying a whole school of fish circling our stationary vessel... and we move on.

By some miracle Świechowski managed to save our last watch from getting soaked and even wound it regularly all through the hurricane. With this, he is able to make a couple of observations and determine our position.

After the hurricane

It appears that the hurricane blew us about 100 km southeast. So we have to think about the best direction for us now. We decide we must head towards the nearest land, and that should be Bermuda, about 700 km away.

The state of the sea falls from 8 to 7; the wind we estimate at Beaufort 7 though some squalls reach 9, but given our truncated mast and, even if we do get beautiful weather, our inability to hoist a full sail, the only way we can gain some speed is precisely with such a strong wind.

Given the conditions we spend a relatively peaceful night, so next morning we immediately get to work. We hoist our mainsail, adjusted to our diminished mast. Then we set to drying out our belongings. Some things—clothing, bedding, books and food—we spread out on deck. Unfortunately, the things were so soaked for so long in salt water that we no sooner put them back into the cabin then hygroscopic law dictates what it must do... everything once again absorbs moisture.

We feel carefree again. For a few days we are covering 90, 60, 70 miles. The sea is calming down. 2-metre waves, a breeze at Beaufort 5 brightens up our seascape.

Finally, on August 25, after careful observation and calculation, Świechowski tells us that around eight in the evening we will see the lighthouse at Gibbs Hill, Bermuda.

We are overjoyed. Two weeks ago was the last time we received the time signal from the Eiffel Tower in Paris; during the hurricane our radio was underwater. Since our watch had not been set for over two weeks it was quite likely that our location could be incorrect. If our watch is off by more than a minute, we'll never find Bermuda.

By eight o'clock we're in darkness. We are looking all around, and see nothing. We consider all possibilities. We now dread that we will be traipsing all the way to the United States with our broken mast...

All of a sudden, just like in storybooks, Świechowski, who had climbed to the top of the mast, shouted: Lighthouse! And sure enough, within a minute of so, we can see the blinking of the light even from the deck.

The wind rises and gets stronger, the waves get bigger, various signs that we are getting closer. In any case, we still have to spend the night on the open sea. We are drifting around 15 miles away and awaiting daylight.

In the morning the weather is getting worse. We lower the mainsail and using just the small foresail we struggle along the waves until dawn.

At daybreak, we raise the mainsail and sail a bit closer to shore.

It is a high and rocky coast covered with luxuriant, tropical greenery, with a lot of sections of some kind of special dark green cedar. Pure white rooftops stand out against this background. All of it so beautiful and alluring. But we know nothing about these islands. All we have is a general map of the Atlantic on which Bermuda is nothing but a dot. According to texts on navigation—because we didn't have maps—these islands are the peaks of rocky underwater formations. Not a pleasant prospect... crashing on rocks just before we finally arrive at a harbour.

Drying out after the hurricane

We decide to circle the island to reach the inside lagoon sheltering the main port, Hamilton.

At around 10 am, the wind has died down, thank God,

giving us just enough to move us along slowly. Sheltered by the island, the sea here is completely smooth, like a mirror. The sun is high, and the water is so clear that you can see to a depth of 20 metres. Unfortunately, we don't at this point understand the significance of what we see. In some places the water is light blue, but in others it is completely dark.

We assume the light sections are shallows so we steer towards the dark until, after careful observations we notice that the light sections reflect innocent white sand lying deep beneath the water's surface, while the dark are sharp formations of coral not far beneath the surface. We throw a lead to check the depth but moments later we are left with just a bit of it, the line having been cut clean as though with a razor.

We are a now about two miles away from shore, among an increasing number of rocks with a decreasing depth, in some cases their peaks jutting out of the water. It is truly a labyrinth. Zigzagging slowly, often turning a sharp right angle we move forward with great difficulty. Then, a slight shock. We are aground. We move swiftly to the bow in order to move the centre of gravity; the yacht, leaning on its keel makes a few indecisive moves then finally gets off the reef.

After two hours of these acrobatics among the rocks and once more resting on a reef so solidly that we had to jump out of the boat, lift and shove then jump back in, we finally make it close to shore.

A cove. A couple of boats are lying upside down in the sand. A couple of black children are splashing around on shore. Unfortunately for us, it is too shallow; we have to turn around and keep going parallel to shore.

At last we meet someone, a black man, a fisherman in a very small sailboat. We wave to him, signalling to come closer.

It took us a while to explain our situation. He gave us directions to continue for a while parallel to shore, then turn between two islands and one hour after that enter a small bay near the town of Somerset. There, after 42 days at sea we finally let go anchor.

Losses from the hurricane as recorded in our log:

- Broken mast
- Broken spinnaker-boom
- Broken tiller
- Crushed superstructure, collapsed and bent handrail and torn canvas
- Damaged skylight, with three broken panes
- 4 deck cleats ripped off
- Cracked bowsprit; cracked gaff, broken handrails in the hatch
- Cracked bow hatch
- Broken door in cabin
- Broken bilge pump
- Hand pump bent
- Two tarpaulins torn off and sunk
- Handrails loosened
- Bent tiller
- Torn mainsail and foresail
- Lost floating anchor with 90 m of heavy 30 mm line
- Lost 100 m hawser 18 mm thickness
- Smashed mast socket.

Endnotes

4. 18th century dramatist, poet, linguist, etymologist and theatrical reformer who was one of the principal playwrights of the Polish Enlightenment. Also a Jesuit.

8

New York Delayed

We are now about 50 metres from land where a large crowd of onlookers is gathering. They are dark-skinned.

Half an hour later a small boat comes to meet us, in it a local doctor, Dr. Sweeney, who offers us his professional certification. He is accompanied by a young man in a white suit, and with what seems to be a clerical collar.

Looking up at me, he speaks in French: "I note you have a religious medal around your neck. Are you Catholics?"

I explain to him that most people in Poland are, indeed, Catholics. "I suppose you don't have any like us here?" I add, smiling broadly in my effort to be amiable with the young cleric.

"You are wrong, sir," he replies pleasantly, "A lot has changed recently. At present about 12 per cent of the population is Catholic. In any case, I am a local parish priest," he adds with a smile.

So, along the way we met the local parish priest... A cliché. How many times have I heard this, how many times have I myself said this when recounting stories about visits to distant friends, while on manoeuvres, on hunting expeditions.

Reminds me so much of Poland. The local doctor as well... maybe he will follow tradition and suggest a game of whist?

Before long I have another encounter, so reminiscent of Poland, and for which there is also a saying, that when you're hungry something will come along. And so it is. All of a sudden a short, dark, elegant gentleman appears and speaks to me in Polish: "I understand, sir, that you are in need of a loan?" Then, noting my hesitation, he waves his manicured hand and continues, "My dear sir, I have here, on these islands, three stores; here, while we are far from home, we must help one another. How much do you need?" And with that, he pulled out his cheque book.

Dear, dear Mr. Machower, if only all Poles had this generosity of spirit towards their compatriots as you do!

A few hours later, a motorboat from the Royal Yacht Club arrives and tows us to Hamilton where the entire executive of the club, including the Commodore, are waiting on the pier to greet us.

After a very warm welcoming reception in the club house, they parcel us out: *Dal* is put in the care of the shipyard; my two colleagues are taken in hand by Mr. Tucker, the local director of the Seaman's Mission; while Commander Brock, of the Royal Navy, takes me under his care.

The first week is spent entirely just resting. The yacht is put, more or less, in order: a new mast is installed and the rest is to some extent restored with the aid of nails, putty and paint.

Taking advantage of free time, we explore the island. On the second day, Commander Brock shows me the devastation caused by the hurricane. Though the eye of the storm passed nearly 300 kilometres away from Bermuda the wind and waves were so powerful that they washed away all the sand from the beaches, leaving bare the rocks that are usually a couple of metres below the surface. A steamship moored at the outer pier with three anchors and with engines going full ahead, was nevertheless dragged around the bay until she was aground on the rocks. On another note, an interesting observation: some trees turned completely yellow from the wind.

The news from the United States is grim: the damage on the New Jersey shore reached millions of dollars. Several steamships were sunk, among them a Canadian schooner, its crew rescued and brought to Bermuda with nothing but the clothes they were wearing.

Because of all this, our survival of the hurricane is considered a great feat, and much more celebrated than just an ordinary arrival by a small boat from Europe. We are stopped on the street, asked for autographs and to pose for photographs. We are asked to tell our story a thousand times—which country are we from, and where is Poland? Etc. etc.

I wanted to know just how close we had been to the centre of the hurricane. To satisfy my curiosity, the Bermuda Meteorological Bureau gave me a map showing the exact route the hurricane had taken and, based on our ship's log, they also calculated the route taken by *Dal*.

One week later, after loading our boat with provisions and water, we are once again on our way, September 3. We are helped immeasurably by the generosity of the Lines brothers who tow us through the reefs taking advantage of high tides and no winds, and send us on our way...

Life is a grand game of bridge—sometimes you get the cards, sometimes you don't.

We no sooner leave the island after just a week's rest and sail off for the American coast when we once again are dealt an awful hand. At the start, not a whiff of a breeze, but when a wind did rise, as if out of spite, it blows straight from the American continent. There is no way to hold our course so we tack into the wind, zigzagging against the wind. Once again the waves wash over the deck. And seep through the tattered canvas on the superstructure.

It's leaking into the cabin. Everything is wet. The cold is getting to us.

Two low-lying clouds hover above us.

For the first time we see whirlwinds, like sea tornadoes. The lower part of the cloud lengthens upwards, as though to connect it to the heavens.

Within a short time we spot three such whirlwinds. Luckily, they are quite far from us, about 7-8 miles. One of them starts approaching us. We read in books that large ships shoot at these winds with cannons, the vibrations of the shots break up the trumpets of wind which results in rainfall.

We are going to try this. But will a rifle shot be enough?

We haul out our rusty rifle but before we can try this the column of water moves away, and then disappears.

The wave becomes shorter, the yacht can't settle on it. While the front part of the hull pitches, the bow dips into it.

My health seems to be failing me. I am out of breath. I know what the problem is; it's my heart.

And so I have an attack, lose consciousness. I'm atoning for my sins: irrational swimming course, some crazy skiing adventures in the Tatras, inter-squadron competitions in the Baranowicz Brigade... Yes! One must pay for everything.

Taking into consideration the state of my health, the pathetic state of our yacht, as well as the discouraging atmospheric conditions (after five days we have covered merely 190 miles), we decide to return to the island.

The weather remains dreadful, the barometer stays low and shows no desire to rise; it is foggy, either strong wind or sharp rain showers mercilessly hit our faces. On top of that we are retreating, giving up!

That is just too much for Witkowski's ambitious nature. Świechowski and I had already noticed at various times during our misadventures that his nervous state was aggravated. But our latest setback—a return to the island—is the straw that breaks him.

I am with him, at this time, in the cabin. Świechowski is above, soaked every minute with wave after wave, fighting alone for our survival while I lie on the cot feeling very weak after my heart attack. Witkowski, usually very glum and quiet, is talking, mumbling in a low voice. Suddenly I realize with horror he's not talking to me but to someone else, and this "someone else" is sitting in the corner in the cot opposite. Witkowski is looking in that direction, talking, and waving his arms.

He is waving his arms more and more, and shouting so loud that Świechowski, without letting go of the helm sticks his head in the opening of the hatch to ask: "What's going on?" I haven't the courage to tell him the truth so I don't reply but I can see that he figured it out with just one glance. We both turn away, averting our gaze for as long as possible. We don't want to face reality.

But of course we have to. On my call, Świechowski lets go the tiller and rushes in to help me. We have to restrain the poor fellow.

The yacht, with full sails and no one at the helm, tosses and turns and leans in every direction, while in the cabin unfold scenes straight from Dante; very hard to write them down.

I prefer not to remember them.

So it is in this sad state that we return to Bermuda. We enter from the north, using the passage that great steamships use, beyond the reefs and marked with buoys. Unfortunately it's night and the buoys are not illuminated. To avoid dangerous accidents, since this area is full of submerged rocks, we cast anchor at the outer pier next to a large coal ship and we go to sleep. During the night the wind changes direction and starts blowing in from the open sea. The boat is rocking again. I can't sleep because of pain in my heart. So I get up from my cot and go up to see if everything is okay on deck. Immediately something strikes me. We anchored very close to the big ship; now its large silhouette looms far in the distance. Our anchor obviously is dragging in the loose sand below.

We move as fast as we can, letting go another anchor overboard, though it doesn't help very much.

We drift more slowly but quite possibly towards some rocks.

At last it's daylight, cloudy, foggy, but daylight. As soon as we get a chance to look around we raise our sails and get away from this unfriendly environment.

In the morning Witkowski wakes up, remembers nothing from the previous night and feels just fine. Świechowski and I are delighted and convince ourselves that the crisis is over. Far from it. The mental state of the previous night went much more deeply than we suspected.

A few months after his return to Poland from our voyage, Witkowski tragically ends his own life. Such a brave sailor; he deserved a better fate.

We return to Somerset where the kind Father Martin insists that I be taken to the rectory. Dr. Sweeney arrives immediately to check me out and orders a long and complete rest. It is November 11.

Before the week is out, a hurricane hit the area between Bermuda and the American coast.

Our situation is not a happy one. I am sick, our funds have almost completely run out. The hurricane season is in full swing. When that ends, then the winter storms will arrive to the US coast, the canals and the Great Lakes will be closed for sailboat navigation because of ice. So after a few brief discussions, our plans take form: Witkowski returns home; Świechowski and I are determined that, come what may, we will spend the winter months in Bermuda and then continue our voyage to America in the spring.

And so began our sojourn in Bermuda.

9

Winter in Bermuda

Our stay in Bermuda remains as one of the fondest memories of my life. Indeed, I can add that it is one of "our" fondest memories because I am certain that Świechowski shares my opinion. The entire population of Bermuda, from the highest to the lowest classes treated us with such sincere warmth and friendliness that one simply must acknowledge this.

One characteristic of Anglo-Saxons is that as a rule they stay aloof from people they don't know. I met a number of "foreign" families who had lived in Bermuda for quite a long time but they lived on the margins, they were not included in the mainstream social life of the island. On the other hand, we are not "foreigners" but sportsmen, and as such we are given a special status. So from the start, all doors are open to us.

Imagine my unease when, after only a couple of days, I am invited to dinner at the home of one of the most prominent people. Everything we own in the line of clothing is totally ruined and absolutely not suitable to wear. I feel like a pathetic creature and want to hide underground, but have no choice but to sit down at an elegant table in an old sweater, torn denim trousers with patches sewn on with sail yarn, wearing old tennis shoes with holes in them and no socks, while on either side of me sat ladies in elegant décolletage. Apart from my sorry wardrobe, my entire appearance is far from attractive with sunburnt, peeling skin and a bruised and battered face. To add to my discomfiture, my knowledge of English is limited to just a few words and phrases. And yet, by the end of dinner, thanks to my genial hosts, with their sincerity and kindness, I feel quite at home.

To save readers the trouble of checking an encyclopaedia, I will briefly describe Bermuda to them.

Situated 1300 kilometres from New York, on the other side of the Gulf Stream, it is farther from Europe than from New York—if you measure with a string on a globe and not on a flat map.

Bermuda is a small archipelago arising from coral reefs, comprising more or less about 350 islands, some of them tiny. I say "more or less" because it's hard to say whether something is considered an island or just a rock. The overall area covers 10 square miles, a row of four main islands accounting for 80 percent of that. The islands are so close together that they are joined by bridges. There are no rivers in Bermuda; the underground water is salty as are the many ponds, which, despite being quite far from the shore are still linked to the tides. Drinking water for the inhabitants is rainwater gathered on their flat, stone rooftops and then flowing in gutters to cisterns under every house. These low, snowy white roofs stand out among the dark green cedars that surround them, occasionally brightened by light and lush tropical greenery – a distinctive characteristic of the Bermuda landscape.

The population is about 30,000 souls of whom the majority, of African origin, is treated as second class-citizens, subordinate labourers. They are prohibited from entering a hotel or a restaurant and have a special, separate section in the cinema.

In Poland, when people speak of Africans, they think in terms of masses of dark, undifferentiated people, so that it's hard to tell them apart. To someone just arriving in Africa this may seem so, in part because there is not much mixing of various races and in part because, to someone unfamiliar with dark faces, a foreigner does not see individual characteristics.

In Bermuda, the "coloured" (that is the official term) are not at all the same since their ancestors were brought as slaves from many different countries of Africa. So you may meet a tall, long-limbed Senegalese or a short, stocky Nigerian. Due to mixing with various Europeans, the skin colour can range across the entire spectrum from black to white. In some cases, a mixed race results in skin colour hardly different from that of a European, although there may be secondary differences such as very curly hair and some facial features. However, the smallest trace of African heritage identifies the person as African and they are rejected in white society. Their position is the most difficult because, rejected by both Europeans and Africans, they are often considered to have the faults of both races.

When the Spaniards discovered Bermuda in the 16th century, it was unpopulated. Despite that, one frequently hears references to Africans as "natives." These Blacks are descended from the sailors in the first Spanish ships that arrived on these islands. Most of them are of mixed race including Indian, Caribbean and Hispanic.

This mixture created a very splendid people, tall, well built, attractive features and highly intelligent. However, the lazy and carefree character typical of the south prevents them from finding work other than physical labour or fishing. In recent years a considerable number of Portuguese have arrived, primarily from the Azores. They are engaged primarily in fishing and agriculture though some have opened small shops. They own most of the fields of lilies and onions that are exported to New York. They are quite distinctive, different from the owners of the islands, the Anglo-Saxons.

There are no snakes or poisonous lizards in Bermuda. On the other hand, when the first English colonists arrived the islands had thousands of small birds in various brilliant colours—bright red, blue, green... Unfortunately, 70 years ago, a British officer longing for Bermuda to "be just like England" imported robins that quickly multiplied and to a large extent took over the habitat of the beautiful multi-coloured original avian natives.

I myself witnessed an attack by a flock of robins on a nest of bright red birds in some roadside shrubs.

In the last couple of decades, because of the improvement in steamship travel, which really has brought Bermuda closer to New York, the islands have changed from being a minor little colony to an elegant holiday destination for New Yorkers. At the present time, everything on the islands is geared to tourism. During the season one can meet famous boxers, film stars, and even gangsters, the underworld royalty of the larger cities in the United States.

Towards the end of our stay on the islands we met a team of marine researchers on an expedition to study the depths of the sea, headed by a Professor Beebe. I had the pleasure of meeting him and his team. Much to my regret they did not begin their deep sea exploration until summer and by then we had to leave and so I never had the chance to see them lowered on a steel line to a depth of 100 metres encased in a round metal cabin so they could observe and photograph the strange and exotic creatures of the sea, both fauna and flora.

We also had a chance to meet some interesting yachtsmen, officers of the British Navy who arrived on their 20-metre yacht, the Tai-mo-shan. They were en route from Hong Kong to England.

Their yacht was built for them, at cost, in the navy docks, and in order to justify their monthly expenses to cover gas and provisions, His Majesty's admiralty assigned them some sort of more or less fictional hydrographic research. We later learned that their monthly budget exceeded the entire cost of our *Dal*.

And you wonder why the British and the Americans set most of the world's records?

We also met an interesting trio—a 75-year old doctor, his 50-year old nurse, and a 20-year old youth—who arrived in their 14-metre ketch. We got to know them and will return to their story later, in America, where we met them again.

A glance at Bermuda right now shows a lovely island dotted with luxurious inns and hotels at full occupancy during the tourist season. They are American tourists and their presence doubles the white population giving the impression that the original Bermuda has disappeared. However, if one wanders around the archipelago, one encounters the most interesting people living enchanting lives.

For example, there is a Mr. O'Connor, with whom I've recently become acquainted. He is the keeper of the lighthouse at St. David's. First time I meet him he invites me to visit him: "Just find yourself a boat, and come to St. David's, once there ask any of the better behaved white kids to show you the way to Daddy's place; you'll find me." Indeed I do. One of eleven of his descendants guides me to the venerable patriarch, the most important man on this island. There I discover many interesting things because St. David's, with no bridge connecting it to any other island, has retained a very distinctive character.

In a quiet bay there I once found a canoe, 15 metres long and only two metres wide, that at one time had been used for whale hunting. However, whales disappeared from these waters about 20 years ago so the canoe lies abandoned, quietly rotting in the shade of cedar trees. On a clear day, one can see on some of the small neighbouring islands the crumbling ruins of old fortifications and their cannons at the edge of the coral reefs rising from considerable depths.

The local aristocracy, descendants of the early colonists, are all related, much like our old aristocracy in Lithuania. Hunched, eagle-nosed and eagle-eyed, often with protruding ears, they are different in appearance from the later arrivals. They don't care for the Americans very much nor, for that matter, the English. They are proud of their pirate origins despite the fact that, all things considered, their ancestors were not engaged in honourable work, using lights to lure ships into the reef, then murdering the crew and stealing the cargo. To this day people recount legends about the great treasures that were buried on various little islands in those exciting times. Supposedly, only a couple of years ago someone found a treasure chest containing many ducats buried in the sand among the rocks. Is there any truth to this? Who knows? The lucky man who found it had no intention of sharing the news, still less paying one-third of the treasure to Bermuda.

One still frequently finds the brass hardware from old ships, the hilt of an old sabre used when boarding a ship, old rock pistols, lead musket bullets, and things of that sort.

In times past, sailors did not have the comforts we enjoy in our good ports. Not long ago a very old African woman recalled a Spanish captain who wanted to scrape and paint his frigate beached on one of the fortified islands. The crew was forced to unload cannons, mast, all kinds of supplies, and then drag the frigate on blocks to the shore. Under these primitive conditions, the work required several months of hard labour.

One day I accompany Commander Brock to St. George's Island. He had to attend to a few matters there so I take advantage of that to look around.

The town of St. George's is the former capital of the island though it is not at all like Hamilton. A small port, a small square bordered with small houses that are a couple of hundred years old. Every stone has a story about the good old days when large sailing vessels arrived bringing rum, sugar, and "mahogany"—which is what they called African slaves. We entered an old pub, the Captain Kidd, and sat down at the bar.

I order a strange mix of rum, lemon, mint leaves and some sort of soda water all poured on ice. The commander, as befits every respectable Englishman, drinks nothing but whiskey.

On the walls of the dimly lit pub are murals depicting scenes from the life of Captain Kidd, the legendary king of the pirates.

Quiet, dark, fresh, it's a pleasant atmosphere. An old maritime wall clock chimes the passing hours.

I close my eyes and rock on my high bar stool. Just then I hear a step on the creaking cedar floor and a deep basso profundo says: Hallo!

“Hello” replies the Commander.

“Hello,” I add to the chorus.

“Whiskey, sir?” asks the black barman, with a wide grin.

We introduce ourselves. He is Mayer, the mayor of the town and honorary consul of Italy, all in one.

He is very pleased to discover who I am.

“You know something, sir,” he says, “my family has many old connections with Poland. Generations of my ancestors once sailed for the Polish kings in the Baltic but then when the Prussians occupied Gdansk, my great grandfather didn’t want to be in their service so he gathered up his wife and children into his sailboat and the family moved here!”

With that he stands, tugs on his grey moustache, clicks his heels, and raises his glass; realizing he is about to propose a toast, we both stand up. “To His Majesty, the King of Poland!” he says with great emotion.

Lord, Lord, where have the past 150 years gone?

At first, Świechowski and I live on the yacht but after a while we realize that in the long run, this will not be ideal. Our sails, ropes and even mattresses and bed linen had been so soaked through during our voyage that they are getting mouldy. After quite a few fruitless searches we finally find what we need.

In the most distant headland of the main island we find a small fishing bay. We manage to secure a 100-kilo anchor and anchor *Dal* very securely in the middle of the bay. With *Dal* protected on all sides by high cliffs, we have nothing to worry about. We transfer all our possessions to our new residence situated at the top of the neighbouring hill.

Our little house, covered with vines thick with red flowers, stands not far from an enormous rubber tree, its branches spreading over the entire area. As a child, that's how I imagined "Uncle Tom's Cabin," made of rough boards and with a corrugated metal roof on which the sound of the falling rain sounds like a marching army and is enough to wake the dead. On sunny days the heat from the metal roof forces us out into the shade of the surrounding trees. Nobody wanted to live in this house so we are able to rent it for the fabulous price of three pounds per month, the amount it would cost daily in a hotel during the season. When I say nobody lives in it, I mean no humans, because there are plenty of other occupants already sharing the place. Spiders the size of a child's hand are crawling across the ceiling; we never get entirely comfortable with them, right to the end. The floor is crawling with ants so to regain control of our table and cupboard where we store our provisions, we set the legs in saucers of kerosene.

On the first night, something heavy jumps through the window and onto my bed. I leap up and see before me a large tomcat. Quite offended that his place is taken but realizing I have no intention of giving in to him he accepts this fact and from then on we sleep together. Our cohabitation has its benefits, confining the mice to the kitchen. Unfortunately, in the spring Mr. Cat gets a very self-satisfied air about him and instead of coming home at a reasonable hour, around midnight, after his hunt, he stays out until dawn. He is now covered with fleas so I suggested that he sleep on the old sails, with which he agreed. Overall, he has a good disposition.

But I mustn't forget to describe the layout of our house: a small kitchen, bathroom with a tub, a separate toilet, dining room, and the "salon," which I've converted into my bedroom. There's a piano, its sound bringing back fond memories of the pianos in the manor houses of old *Kresy*[5]. There's a desk by the window, which is where I write most of my memoir of our voyage. It's a large window; it frames a big stand of old cedars, over the tops of these one can see other trees growing on the rocky slope, and in the distance the boundless sea, sometimes agitated and sometimes as smooth as the pools at the Łazienki Palace [in Warsaw]. In good weather, with a light, moist breeze in the air, when the tide is out one can see waves of foam on the sea, far, far on the horizon, about 15 miles from shore; depending on the wind, one can see the entire crest of foam, sometimes broken up, and sometimes not at all. That's where the reef is, the very reefs we got stuck on when we first approached Bermuda.

On early maps made by the first sailors who came here, one can see that where these reefs are now there once was an island, an island of impressive size, with trees on it. The waves, the winds and the rains did what they must do. All that remains is a lonely, bare rock. Actually, the entire archipelago was once much larger. Evidence of this was discovered during the digging for the canal to enable the entry of steamships into the lagoon when not only rocks were removed but also great quantities of cedar trunks preserved for centuries by the salt water and resin.

The winter months pass quickly. I have a bicycle, which I use to get our provisions from town, and to ride around all the islands to visit my many new friends. Świechowski, meanwhile, prowls around our nearby neighbourhoods. Frequently upon my return from town unexpectedly early, I find him sitting in the shade of the trees near our house repairing our sails in the company of one or two ladies who live close by, and who provide the motivation for this pastime.

In the fall, taking advantage of the ebb and flood of the tide, the two of us along with the help of one African, bring our yacht up on the shallows, and scrape it clean. While we are at it, we also paint it to prevent damage by foul weather, water and teredos during the winter months.

At least once a week we go fishing. Świechowski's favourite pastime.

In beautiful weather, we take a motorboat to the outer reefs and drop anchor somewhere over these submerged rocks. Then, leaning over the side of the boat with a long fishing line we gaze through the clear blue water looking for our victims.

Most of the time these victims don't keep us waiting long. After only a few moments I hear somewhere behind me, the sudden thrashing of a tail against the floor of a boat, a fish caught by some luckier, or perhaps more accomplished fisherman. But after a few hours the bottom of our boat is filled with a great variety of brilliantly coloured fish, some of them a metre or more in length.

The most common edible fish in Bermuda are: "blue fish," "red fish," silver bonito, and brown rockfish. We often encounter a small species of shark about a meter-and-a-half in length. The only people eating these are the Africans who boil them for over ten hours, seasoned with some kind of very spicy Creole seasonings.

When at last after a long fight you are hauling in this "little fish," you have to crush his head with a powerful swing of an axe. One of my friends failed to do this and had his leg broken by a powerful swing of the shark's tail.

Sometimes a school of these fish will surround a boat, scaring away all other kinds of fish. Fishermen deal with this situation this way: instead of killing it and pulling it into the boat they execute a clean cut along its side from which then emerges a liver of gigantic proportions; they then release it from their hook. This bleeding monster uses all his might to escape while the rest of the sharks, like a lynching party, chase him to the depths.

But the fishermen's most dangerous enemy is the green Moray eel. They are two or more metres in length, with an enormous mouth and monstrous teeth; the Moray can crunch a lobster as if it was a mere beetle. This one too is extremely dangerous to haul into the boat. A bite from a Moray is serious.

Sometimes we sit on old wrecks to look for lobsters, which are nothing special here. For example, one can get a beautiful big lobster for—in our currency—3-4 of our *złoty*.

So we pass the time in this lovely environment. One day we receive a flag and a splendid album of photographs of our harbours, our yachts, and our activities at the Officers' Yacht Club in Gdynia.

During an official visit to the Royal Yacht Club in Bermuda I presented the flag and the album to the Commodore, along with a letter from Commodore Spałek expressing his gratitude for the warm welcome we received on these islands.

Commodore Goslin, who received me, in turn presented me with the flag of the Royal Yacht Club along with a letter addressed to Commodore Spałek.

From the letters and cards we are constantly receiving—from cabinet ministers, the general public, and every high school student in Poland—it seems everybody in the country is wishing us success.

That makes us realize that they are not just acknowledging our voyage but counting on us. There is no way we can quit.

Therefore, with the coming of spring we pull our yacht out on land and get to work with renovations. The rudder axle was totally ruined and has to be replaced. We also decide to add some 700 pounds of lead to the keel to stabilize the boat more. Then on to reinforce the superstructure, add a coat of paint, brand new sails—a gift from the Officers' Yacht Club of the Republic of Poland—some new rigging, some carpentry repairs, load provisions for a three-week journey, a couple of hundred litres of water, some maps and... voilà, we are ready to go.

The time has come to bid our farewells, which makes us realize how difficult it will be to leave these islands behind. We realize, of course, that we will never be able to return, and yet it is so difficult to come to terms with the reality that we must leave so many people who have become so dear to us.

Dick Tucker, the director of the Seamen's Mission, our most faithful friend; Sir Watlington, the mayor of Hamilton and his family; Mr. Pursel, the editor of the "Royal Daily Colonist," Mr. and Mrs. Davis, Mr. Machower, Dr. Dickenson and so many, many others.

After many farewell parties of all sorts ranging from formally official ones to others including those in our own "palace" where the menu invariably consisted of borsch, lobsters and generous helpings of rum, the time has come to leave.

On June 3 in the morning, a large group of our friends gathers on the pontoon pier where *Dal* is moored, gently rocking side to side.

Everybody is here: the local island aristocracy; American tourists; the French artist Verpilleux, who is a friend of our [Polish artist] Tadeusz Styka; and the conductor of an orchestra in one of the grand hotels; a retired Russian officer; several British naval officers with their wives (one of them a Finn from Helsingfors with whom we have several mutual friends in Finland); but the largest group by far are the Portuguese fishermen and the Africans, all of them friendly and kind and it was obvious to us that they were sorry to see us go. Many of them bring us parting gifts—something to eat, to read, a scarf, or something of that sort. "After all, you are going to a cold country!"

At exactly eleven o'clock, we cast off the mooring lines and sail off. We must get past the outer reef while the tide is high so there is no time to waste. We hear our friends shouting "good-bye" and shout in return ourselves. Before long the pier and all those on it waving their hands and handkerchiefs disappear behind a rocky little island. I feel a tug at my heart, because I will never again see all these warm, kind and friendly people. Leaning his elbow on the boom that our main sail is wrapped around, Świechowski is controlling the helm with his foot. He is taking care not to deviate from the safe course set for us by *We Two!*, which is kindly towing us.

André with the editor of "The Royal Gazette"

The good *We Two!* So many great excursions on this motorboat! We are approaching the place where we used to set up our campsite. We Two always had a hard time slipping in there at high water. There's a large flat rock there, perfect for warming up in the sun after a swim. We move past some charred rocks, the remains of a campfire, and fantastic cliffs jutting straight up along the shoreline. We sometimes came here with a large group for some swimming in the moonlight. So many memories!

A few miles to the right there is a British naval base. A couple of cruisers at anchor. They raise the drawbridge to let us through. And there's the bay where we first dropped anchor. In time this too disappears. We are heading towards the reef, the tide is high, the surface water is completely smooth. The sun is at its zenith, penetrating the water 30-40 feet. We are once again above the enchanted underwater garden full of extraordinary, mysterious rock formations covered with luxuriant marine vegetation.

Departure from Bermuda, *We Two!* by their side to tow them out of the bay.

The Lines brothers, who had offered to tow us in order that we save about 30 miles had we taken the same route as steamships, have taken on a tough assignment. Far ahead of us one of the brothers is on a small racing boat scoping out the sea for underwater rocks that at this point are not far beneath the surface. He waves to us signalling when we should turn. The towboat gradually slows down; we are now over the reef. We stop a couple of times, make several sharp turns, then, after a few minutes, we are out on the open ocean.

It's time for our final good-bye. All three boats approach one another; we shake hands for the last time, hard and long. They are, after all, islanders; they understand everything. We exchange very few words. Only Hilda Line, the one sister among these four brothers, whispers, "May God guide you," and then we sail away from one another.

We are alone between the ocean and the sky.

Endnotes

5. Once the eastern region of Poland, or borderlands, now divided among Lithuania, Latvia, Belarus and Ukraine.

10

Bermuda–New York

What lies ahead of us now is nothing less than a task for a manta ray.

The greatest part of our voyage to New York is behind us but we still have 700 nautical miles to go. For perspective: about the same distance as Gdynia to London. However, it's not the distance that is daunting but that we have to cut across the Gulf Stream.

Here's what sailing books have to say about the Gulf Stream: A current of warm water surrounded by cold water. As a consequence, there is almost always bad weather around it.

Not a great prospect.

During our extended stay in Bermuda we had the opportunity to see a lot of yachts, and to discuss amongst ourselves and with others the many issues that could arise when sailing in a small boat. Because of this we were able to clearly evaluate our own capabilities and also that of our *Dal*.

We no longer have Witkowski with us so we have to make do with a crew of two, each of us taking two watches. I take the 12 to 6 watch, Świechowski takes 6 to 12. It will be harder, but on the other hand, we have more space in our small cabin.

My health is in good order; Świechowski has just recently recovered from stomach flu, a severe one. He lost about 20 pounds. He feels better but he is very skinny. He insists that everything is okay, that he has never felt better, but in my heart I don't really believe him. His style and character will never allow him to admit to a weakness.

As for our poor little Dal, although it was well and strongly constructed, she always was and will always be a yacht for coastal sailing. While the gaff rigging is very attractive, it is not at all easy to hoist and lower the sails. The keel starts only at the point where the mast is stepped, as is often the case in racing boats. A leaky hatch, a huge skylight, and the worst cockpit—the open space in front of the rudder—is not for just one or two people but could be a promenade for eight people.

Now, more than ever, we are very conscious of these shortcomings. But there's nothing we can do about it. You can't convert a villa into a fortified bunker able to withstand an artillery attack.

Finally, being Poles, we also have to contend with something else, the worst thing of all: ridicule. When the French or the English attempt something and their experiment fails, all is well; they will be praised for their effort, for the courage to try. But if we fail, everyone will laugh at us, especially our own people.

We are sailing along briskly and happily. The sun is shining. Our first night out our dear friend, the Gibbs Hill lighthouse, keeps blinking a farewell at us right until dawn, just as it greeted us eight months ago when we first came upon the island.

The seas are gentle, the deck is dry, I am in the cabin tidying up and putting things in order. Świechowski is at the helm, singing merrily and driving me crazy with the same three stanzas of an old Warsaw hit. He doesn't remember the words and probably never did learn the melody.

The weather up there must be wonderful and the sun has warmed him up because he goes more and more off key. I am thinking blood, murder, the most terrible tortures... I hope a thunderbolt strikes him from the clear blue sky! But I sit quietly and say nothing.

There's a reason for my tolerance.

On the third day of the hurricane there was a moment that I will remember for as long as I live, even if I were to live a thousand years. My entire skin is dry and chapped but I don't feel it anymore. I don't even feel the ropes that I tied around myself to keep myself from being tossed overboard if a wave were to hit me or tip the boat on its side. It is so very dark and my eyes are burning from the salt, so I am convinced I've gone blind.

We bear this in silence for hours on end.

Then suddenly, in this darkness, through the roar of the hurricane and the fury of the seas, I hear a voice singing off-key: Dulcinea tra-la-la...

That's when I realized I was still alive.

So I know that Świechowski will keep on singing right until sunset.

The wind is dying down. The sea is as smooth as a town square. Every now and then we feel a slight breeze, we move a few metres, then stop again. From time to time an underwater swell barely disturbing the surface gently rocks the yacht and lightly jerks the limp sail and braiding it into the rigging. Eventually the wind stops completely. There's a deathlike stillness all around us except for two or three bonitos (small tunas) circling round us in the clear green sea.

During this watch Świechowski comes up with an extraordinary way to catch fish, but since a kitchen knife attached to a stick yielded poor results, he improvised a novel system: he removed a hook from a clothes hanger, straightened it a bit, filed it to a sharp point, attached it to a rope the thickness of a finger and, with a lot of bacon on it, tossed it overboard. The bonitos promptly disappear. Perhaps for several reasons. Maybe the hook frightened them. I can understand this. It's not often bonitos encounter a coat hook with bacon attached to it. Most unfortunate. But there are other reasons too: small fish swim away when bigger fish come close.

The morning sun gets hotter, the fish aren't biting, so our fisherman decides to take a nap. But to make sure he is awakened in the event of activity on his fishing line, he ties it to his leg.

As he recounted later, he was awakened by a light tug, then a stronger one, then no two ways about it: he hooked a fish! He pulls on the line but at exactly the same time as the fish pulls on it too.

I wake up with a start to the sudden strong rocking of the boat and the sounds of a struggle on deck. "Quick, quick!" Świechowski yells. By the time I get there, the struggle is over. There stands my friend with an expression of unconcealed pleasure on his face, in one hand a large kitchen knife, in the other the cut end of the rope.

Later he regaled me with stories about the unbelievable size of "his" fish. As a rule I don't believe fishermen's tales but in this case there is one fact to which I can attest: the skin on his leg was shredded.

After that we don't fish any more.

July 6, at the break of day, the wind starts slowly rising. Good weather still, sunny. Around 2-3 in the afternoon we see some sperm whales. Hard to say how many—two, three, more?—because they surface only for a few seconds and disappear again.

Just as my watch is ending, at sundown, I call down to Świechowski to get up on deck as quickly as possible. Just a few metres away a shiny black mass, about the size of our yacht, appears on the surface. We hear a low, hoarse murmuring sound, a small eye blinks at us, a column of vapour shoots straight up into the air.

It's unsettling to see something like this just before bedtime.

We have decided that, as a precaution, we should reduce the surface area of our sail almost by half for the night. After dark it is hard to see the surface seas to help evaluate the wind. In any case, despite our precautions the Gulf Stream gives us an unpleasant surprise. Once in its waters we are startled by a sudden squall.

We have to deviate off course at racing speed for about 20 minutes close-hauled to prevent our sails getting torn.

Next day at dawn we notice that water is no longer clear and very different, with areas of floating seaweed here and there. So, we must be on the alert for the next 300 miles!

Bad weather did not keep us waiting for long. Bad weather? Well, on land such weather would be absolutely fine; the barometer didn't drop... in fact went up a bit. During the day the sun was hot as hell, at night the sky was covered with stars. But none of this interfered with the wind, which constantly increased in strength.

Once again we faced difficult conditions. We didn't even consider completely furling the sails or paying out the sea anchor because we knew from experience that in these conditions the yacht does not adjust to the wind and waves and secondly, in the Gulf Stream the weather can remain unchanged but the current can still move us God knows where back into the ocean.

We have no choice, therefore, but to propel ourselves forward as much as possible.

We keep the sails furled the whole time yet despite that our speed was exceptional. We covered almost 150 miles each day despite high, crested seas that washed over the deck non-stop. It is once again a battle for survival against the elements. Controlling *Dal*'s direction requires total attention. The waves are constantly throwing the bow off course requiring that as soon as we break through the crest of each wave we have to exert all our strength to straighten her out.

The light in our compass is not functioning so we have to steer according to the direction of the wind using a torch to check the compass from time to time.

The next day we are so tired that it is too hard for us to sit at the helm throughout an entire 6-hour watch without a break. Sleep deprivation, nervous tension, and mental fatigue are powerful enemies that can occasionally defeat us.

The third night of our race against this raging sea stands out in my memory as though it happened yesterday. I am standing at the helm attached with ropes to both sides of the boat lest I lose my balance. I feel I can't risk sitting down because I could instantly fall asleep. In my delirium I can hear people talking, and some repetitive jazz melodies. I shudder. The wind, salt water and sleeplessness burns my eyes. Forcing myself to open my eyes as wide as possible I stare ahead, and stare ... Ahead of me I see a faint whiteness, a sail? And some lights? Suddenly I realize that we ourselves have no lights and we are far, far from land, so this must be a ship, its red and green lights showing its position. It is sailing straight at us; I shine our torch on our sails so that they will see us and I call Świechowski to come up in case I need help.

They've turned on their searchlight, and changed course to avoid us. We figure the danger has passed but not at all. I forgot that at this distance from land and in this weather our yacht looks like a lifeboat so they will approach us to rescue us. The ship is manoeuvring around us. First time we face this kind of danger since our experience in the hurricane. Every time we rise on the crest of a wave we are blinded by their searchlights, and moments later when we fell off the wave, we descended into total darkness. For the next hour we are caught in this game of blind man's bluff among the waves. After a while we are able to take advantage of a time when the ship is alongside us and alongside the waves so there is no mountain of water constantly rising between us enabling us to signal with our torchlight: "The yacht Dal from Poland to America. All is well."

We will never know if they saw and decoded our signals or if, given the night and the weather, we simply slipped through their fingers. In any case, we didn't see the ship again.

For the next two days the weather is much calmer but at the end of the second day the barometer once again drops sharply.

Once again we prepare for the worst.

We tie down everything on deck, oars, boathook... we cover the forehatch and the skylight with tarps. We even get the sea anchor ready, but as we get closer to New York we no longer need it.

When I once told this story to a very well known Polish sea captain, he gave me this advice:

"Instead of a barometer, just draw a picture of a donkey on a poster, attach a tail on a string, and take note: if the tail is not moving—you are becalmed; if the tail is wet, it is raining; if the tail is moving, there's a wind; and if the tail gets detached, it's a hurricane."

One night, there's a great glow ahead of us. It's New York.

In the morning the wind has picked up, we are sailing along at a brisk pace. We are passing "the Battery," where the shore is sharply delineated as though cut geometrically, its surface a lawn with a row of old cannon topped with a large American flag. It is fluttering, its red and white stripes and silver stars blending in the sunlight and the morning breeze. For someone unaccustomed to it, sometimes it seems like a very complicated pattern... at other times it is so straightforward and logical in its representation. I am seeing it perhaps for the one-thousandth time but today it makes a particular impression on me because I see it for the first time on American soil.

11

New York

According to local rules, we must first enter a quarantine area where we let go our anchor, lower our sails and hoist the quarantine banner on our mast. We are waved forward from shore. So we weigh anchor and hoist the sails and moor ourselves by the pier in the quarantine dock.

We have barely managed to do this when a photographer and a reporter appear… out of nowhere.

"Stand beside the mast, wave." Click! The first picture is taken.

"Now shake hands. Smile!" So we shake hands and affect a ghastly grimace that bears no resemblance to a smile. Click!

"And now, can you just say a few words…" says the reporter waving his notebook and pencil, but as I am getting our yacht documents in order to register us and am trying to get out of our boat as quickly as possible I can hear Świechowski's monotone: "We left Poland on June 5… etc. etc."

And to think… we will have to repeat this story from beginning to end several thousand times and already it is making us feel nauseated.

So now that everything has been documented we must rush as quickly as possible to our own place, Pier No. 8, the Gdynia-America Line.

Also in quarantine is a large trans-Atlantic liner waiting to attend to the usual formalities. We want to get past it but almost as soon as we get to its bow it moves. Its huge metal wall rises above us and blocks the wind. Our sails go limp. Moving its huge bulk forward the ship creates currents that pull on us, but this is of no consequence to these gentlemen having long ago forgotten that sailboats are dependent on wind in order to manoeuvre. We remain there, cheek-by-jowl, our rigging occasionally catching on one or another protuberance from the ship.

Some lovely smiling passengers on the top deck, not realizing the danger of our predicament, lean over the rail cheerfully trying to hook our flag on their parasols. Some officious fool, his uniform bedecked with gold braid, indignantly shouts at us: "What are you doing here?"

If I could I would happily send this type of guy on a long voyage on a sailing ship carrying saltpetre from Chile to Europe so he could learn something about sailing, his own profession after all.

Idle thoughts. Silently we push ourselves away from the steel wall of the ship with all our strength. To make matters worse, we push ourselves next to a place where water is pumped out by machines in huge quantities onto our deck, which then seeps through the hatch into our cabin. In a matter of seconds the cabin floor is underwater. We safely pass the ship's enormous propellers as they rotate just barely under the surface of the water.

Wind once again fills the sails moving us sharply forward; we bail out the cabin, our anger subsides, we are sailing upriver, slowly because heavy rain slows the wind down. Świechowski is at the helm, while I, having bailed out the water, start telling him some really awful jokes. Świechowski grimaces and seems to be shaking but I don't know if that's laughter or a reaction to the cold, but just in case, I pass him a bottle of rum.

Slowly we sail past one pier after another—and then a big tugboat arrives to meet us. It circles us, we throw out a line and in this fashion we travel the last quarter mile at great speed. Just as we arrive at the dock, as we turn, the hawser snaps. We are a sight to behold, dropping our sails and coming to a stop at the designated place.

And suddenly something terrible happens: a gang of people armed with cameras stream out of the hangar. A Mr. Wasilewski welcomes us in the name of the Consulate General. We shake hands.

Photographers are swarming us from all sides, barking out orders with occasional pauses when we hear the clicking of the cameras; while somewhere on the side there is the whirring sound of a movie camera:

"Stand closer together please, wave your arms in a greeting..."

Flashbulbs go off: "Shake hands!"

Then it's the reporter's turn and I begin my monologue: "We left Gdynia on June the... etc. etc."

"No, No!" one of them shouts and pokes me with his pencil. "That's not what I want. Who is your favourite actress?"

I am at a loss.

"I remember. I met a famous star in Bermuda. And I remember that she was very attractive. But her name? Kill me if you wish... but I just don't know her name. And besides, I have never seen her in a film."

So I stop talking and feel very foolish. At this point I suddenly hear Świechowski—unabashed and confidently replying to all questions. Nothing about him surprises me anymore and today I know:

> The best film actress is Pola Negri.
> The best director—Bolesławski
> The best and sweetest singer—Hanka Walska
> And the prettiest American woman—Miss Chicago, of course, Alicja Jagłowska. (The perfect answer considering our Polish American audience.)

And then what? I must admit all that chaos remains chaotic in my mind. We barely have time to change our clothes when we are taken to some German restaurant. As soon as they realize who we are the orchestra begins to play *Boże coś Polske* (a religious, patriotic song). And then an oberek (a Polish folk dance), then the owner of the restaurant joins us at our table and then... well, the next day I wake up back on the yacht with a monumental headache.

For the next few days we are entertained by a non-stop parade of reporters, photographers, acquaintances, sportsmen and others who are simply curious. Sometimes they want a professional interview for a yachting magazine; next they want a radio interview; then somebody asks us to brush our teeth with a toothpaste they want to advertise while others want to get a picture of us smoking a certain brand of cigarette; still others suggest we put our yacht on exhibit and charge admission.

"You are stupid! Nobody will respect you more, if at all, for giving away what you could get paid for! Take whatever you can get. In a couple of years nobody will remember you but the money will be yours."

But then there are other voices: "Your country does not have the financial resources to promote its interests in America. Don't squander your accomplishment on paid appearances for personal gain, let us take advantage of your publicity, every appearance, every radio interview, every article—that gives us an excellent opportunity to inform Americans that Poland has access to the sea and that Poland has a navy!"

The situation is altogether too much! Nevertheless, we have to decide on some plan of action. Chicago is hosting the World's Fair 1933-34: A Century of Progress. The previous year the Italian Government sponsored General Balbo, together with his squadron [of 24 seaplanes on a round trip from Rome to Chicago]. This year, [1934], Admiral Byrd's ship from his Antarctic Expedition is still on display. Consequently, at a rather tense meeting at the Consulate General [in New York], it was decided that we must go to Chicago with *Dal*. The difficulties immediately start piling up. The easiest way of getting *Dal* to Chicago is by rail, but that would diminish our transatlantic sailing feat considerably. Back to the ocean to Canada then sail up the St. Lawrence against the current to the Great Lakes? That would certainly make for a long, long journey. Fortunately, there's a canal linking Lake Erie to the Hudson River. And, luckily, along this route there are many large Polish communities. So that's the route we'll take. Unfortunately, it is impossible to use sails to pass through the canal. So we have to install an engine. And that leads to the question of money. We have pretty much gone through the funds we had for our expedition. The solution presents itself this way: the Polish communities along our route will collect funds to donate to us to pay for the engine, however the problem with that is that they will give us the money when we get there, i.e. in the future, but we need it now, so for that there is an immediate fundraising drive in New York and New Jersey. My last few dollars also go towards that goal and so... we buy the engine. *Dal* is lifted on land in the shipyard, the engine gets installed.

No doubt all that I have related in this story seems so easy and straightforward but in fact it is a great undertaking with a lot of problems, worries, endless talking, running around...

During these three weeks in New York, Świechowski and I have no time so see anything except for a casual view of the city from our boat, and the occasional invitation to a reception. The sudden change in plans takes over our lives. The crash, bang and general noise of the streets and public places, the subway, and the constant rush in all directions completely throws me off balance. As soon as Janusz Iliński wakes me up before rushing off to his office, he rattles off my instructions for the day in the language of a telegram: "Don't forget! 11:45 at Tom's. Then with Moszkowski for lunch. At two o'clock you attend to the engine. Drop in at the consulate. Buy yourself a couple of shirts because you look like a ragamuffin. Call Mr. and Mrs. X, tell them you have to give a talk or something tonight because we are going to the country. Tomorrow radio, a party at Zosia's, then the Veterans' group, and don't make plans for Saturday or Monday...bye."

I'm not accustomed to this pace. I am dumbfounded, bordering on idiocy; I forget things, confuse things, make commitments with different people in different corners of the city all at the same time, so I collapse in despair and want to die.

A surprise! The Standard Oil Company offers to donate all the gas and oil we will need for the entire voyage to Chicago and since we can't take it all with us, we can fill up as needed along the way at any branch of this company because this company has stations everywhere in the entire United States.

We are also presented with an entire folder of maps of the areas we'll be passing through. What a pleasure it was to study them all. It has been quite a while since I had land maps to look at. The last one was the army map of Baranowicze, "scaled one to a hundred thousand." But with these, instead of easy names such as Stołowicze, Ciukatowicze, Podlisiejki, I see strange places with strange names that are spelled one way but which are pronounced a completely different way.

The yacht is finally "motorized," ready at last. We set off with a few friends for the first test. How strange it was, how awesome to feel *Dal* moving away with its sails furled. I sit by the helm and all I have to do is move the tiller to port, or to starboard, for the boat to obey my command with no concern for the direction of the wind. Świechowski, who had advanced to the lofty position of chief engineer-mechanic frowns and expounds on the complexities of carburetors, pistons, and crankshafts. He speaks so beautifully that he impresses everyone, including me, with his knowledge even though I know that he knows no more than I do about engines and he learned just a little while ago how to start and stop it. However, praise God and the Standard Oil Company, at every station there will be good mechanics to check up on our engine... so we won't fail in our mission.

Our day of departure approaches. Consul Kwapiszewski has been persuaded to accompany us for a couple of days. At the appointed hour he arrives with crates full of provisions, bottles, and a large beach umbrella. If we are to travel in style, then we travel in grand style.

And so on the evening of July 2, exchanging fond farewells with our dear friends, we set off on the next leg of our voyage, and on to our final destination—Chicago.

12

New York–Albany

Although the distance between Chicago and New York is only about 1200 kilometres if measured in a straight line, this distance doubles following the waterways. Our route takes us past seven states. We have to go against the current up the Hudson River, then the full length of the Erie Canal and then along four of the Great Lakes: Erie, St. Clair, Huron and Michigan. We have to hurry. "Polish Day" at the Chicago Exhibition is on September 26 but we must also stop at all the towns along the way that have sizable Polish communities.

Since it is still daylight we deviate off course to at least see the Statue of Liberty at a distance and sail past that part of New York where the gigantic angular skyscrapers reach to the clouds.

As night descends over the city the entire area is illuminated with millions of lights. All those monumental geometric shapes with brightly lit windows arranged in regular rows while multi-coloured neon signs scattered everywhere create a fascinating image. The facades of some buildings have light projected on them creating a milky contrast to the others. Consul Kwapiszewski identifies them: This one... that's

Rockefeller's Radio City; and this... the Empire State Building, the tallest building in the world, taller than the Eiffel Tower by several floors. Later, all these buildings will be largely empty. They contain mainly offices and commercial establishments. Within a couple of hours great hordes of people will spill out of the buildings and go home in their cars, on subways or buses or trains to the suburbs beyond the city.

Around Brooklyn Bridge between five and eight o'clock, underground and above ground trains rush in several directions carrying an unbelievable three million people!

We turn into a small harbour to get some gas. About two hundred metres away we see the George Washington Bridge, the longest, single-span suspension bridge in the world. It won't hold this record much longer; construction is about to start on a similar bridge—but a couple of kilometres longer—in San Francisco.

We leave our little gas station heavily loaded. Not only is the gas tank filled to capacity but we also loaded several small containers in the cabin, between our bunks. Apart from that we also have our supply of oil for the entire voyage. The atmosphere in the cabin now is unbearable. The engine is so noisy that no conversation is possible and the smell of gas is so strong that after every few hours of sleep we wake up with a pounding headache.

Just before we left New York we heard about an accident that befell our good friend, old Dr. E.P. Hussey, the commodore of the Buffalo Yacht Club who was visiting Bermuda on his ketch, and about whom I wrote earlier. On his way home from Bermuda, when he was already on inland waters, a fire broke out in the cabin while he was cleaning its navigation lamp. Despite getting immediate help, the fire destroyed his yacht. I fear that we may share that fate. After all, we are loaded up with gas up to our nostrils! Well, if nothing else, we would treat the Yankees to a great spectacle.

As a precaution, we decide against any cooking with our stove and instead of using our kerosene lamp we use battery operated electric navigation lights. Neither of us ever smokes and we scream bloody murder if anyone with a cigarette approaches our deck.

The night is totally dark. We have left the city behind. We are actually in New Jersey waters staying close by the left bank and see hardly any ships. We take turns at the helm. To the right of us there are speeding trains on a track by the water's edge. We have to steer carefully, while checking the map the whole time. Here we are to turn, immediately after that we must go around a red light on the right... the engine makes a monotonous noise... our eyes are sticky.

Slowly day is dawning; the thick mist over the river rises revealing the majestic landscape. The breadth of the river we're sailing on is five, six times wider than the Vistula. The landscape is vaguely reminiscent of our sub-Carpathian region but the slopes are steeper, sometimes even precipitous, here and there one can see granite. All around us are trees. What an amazing sight this must have been a couple of hundred years ago when the first colonists set forth into the interior of this unknown continent in Indian canoes.

Just around a bend in the river we see a very high and steep forested rise above the riverbank topped with fortifications. West Point.

I have heard and read so much about West Point. This is the military academy—modelled pretty much along the lines of our military colleges—set up long ago by Tadeusz Kościuszko. Never in my life did it occur to me that one day I would see this. The reality exceeds anything I could have imagined.

We stop at first on the other side where we have an appointment to meet someone at eight. We arranged this meeting by correspondence.

Exactly at the appointed hour this "someone" appears. I recognized him immediately from a distance. Nothing about him has changed—perhaps a little gray hair—but as ever the same crisp, elegant silhouette of an old soldier-gentleman.

Colonel Farman![6] Instinctively I stand at attention. We shake hands.

Old memories return, from 8-10 years ago when the Colonel was the American attaché in Warsaw. A couple of weeks ago he noticed my name in a newspaper and immediately sent me a gracious letter of congratulations and an invitation to visit him at West Point. It is because of our renewed acquaintance by correspondence that we have stopped here.

We sail across to the other side of the river. We approach a small harbour, its buildings serving also as a railway station. A couple of workmen come to have a close look at us. Typical Yankees. Thick workmen's boots, denim overalls, shirts with rolled up sleeves revealing tattooed arms.

When we stop at the pier I want to hook a rope over a post: *"A rzuć ta swój śnurek, pomogę ci."* ("So throw yer rope to me; I'll help ya.") I am stupefied. First time we so casually meet a Polish immigrant. *"Jak się masz?"* (How are you?) he adds, heartily taking our hands. He already knows all about us, having read the news in the papers. He immediately recognized our Polish flag. He translates everything in great detail to his co-workers, with the air of a man who knows us.

Unfortunately, we don't have time for a longer chat. We have to catch up with Colonel Farman. He went ahead of us in order to change into his uniform, which, as is the American custom, he wears only at the military academy.

We spend the next several hours touring the school. Unfortunately, the students are all away on holidays. The Kościuszko Monument. At his feet, the chain that he had stretched across the river to block the British flotilla. In the park along the slope leading to the river, Farman shows us the spot where Kościuszko liked to sit.

In the hall of honour, among a number of attractive paintings, there is a splendid portrait of General Casimir Pułaski.

The Commanding Officer of West Point is away; however, as an officer, I feel an obligation to sign his guest book.

We visit the stables, the barracks. The stable hands are exclusively African-American; their cavalry sergeant is black as tar. As I can tell from our conversation, this is a very intelligent man. During the war [WWI] he reached the rank of Captain; after the war, when the army was reduced in size, he was once again demoted to a non-commissioned officer.

We have dinner in the Officer's Mess. There are only a few officers here, the rest are on holidays. It's been a very pleasant visit and we'd have enjoyed staying longer but unfortunately, we are pressed for time. Colonel Farman kindly drives us back to the pier.

The engine starts up. We are on our way.

The panorama unfolding before us is magnificent. Along the shore are scattered houses and cottages partially concealed by trees.

I recall an article I once read in a Polish newspaper, the writer a resident of a Warsaw suburb. He asserted that one can't demand from suburban dwellers that they make their houses attractive; they can't afford it. Beauty, however, is not always connected to money. After all, quite frequently a shop girl who remodels her hat herself every spring is much more stylish than some ladies from high society. It's all a matter of taste and imagination, nothing else.

On the Hudson one encounters glaringly poor taste infrequently and more often than not, among the rich. I saw some houses in the Egyptian style, their plasterboard columns painted with hieroglyphs, an Assyrian palace with a mural on its façade depicting bearded Nebuchadnezzars chasing miniature lions that somehow resemble bulldogs. And whenever one comes upon a house in the "Colossal" style there is no doubt as to who the owners could be. On the other hand, the small houses almost without exception are lovely. Their one flaw is that they are too much alike, but that can't be helped and one must just get used to it. Everything here is produced on an assembly line: shoes, cars, films, clothes, even women who, although beautiful are boring in their sameness. Every day one sees hundreds of Greta Garbos, Bridgid Helms, or Loretta Youngs.

The real plague of the American landscape is the billboards and advertising in every possible size and shape. Never have I seen a country with its scenery so defaced with countless ugly billboards. I came to detest a certain kind of lemonade and chewing gum, even before tasting them, just on the basis of their invasive advertising.

In the afternoon there is something out of order with the engine, an odd knocking sound noticeable even to the untrained ear. We head for a Kayaking Club in Poughkeepsie. We take down our flag to avoid long conversations that could impede the work. We get a mechanic who turns everything upside down in our yacht. He opens something, then he closes it. The entire boat—and we ourselves—is splattered with oil and grease.

We proceed. It's so beautiful that our good humour returns. It's summer holiday time: there are young people in bathing suits everywhere on the shore. We often pass kayaks or various rowboats with a crew of tall, well-developed boys and girls floating lazily with the current. Usually we just wave the traditional greeting and go on but sometimes the kids can instantly tell who we are. In the sporting world there aren't many who aren't up to date on everything.

They get to their oars and although we are moving quite swiftly they soon catch up with us. At the same time, while rowing with all their strength, they still start a conversation.

"Hey, where are you from? Oh, is that you? We know, we know. And how long is your yacht?"

Getting rid of them with a joke is not possible—they smile sincerely, reply with good humour, but don't leave you until they've got all the answers they want. Out of breath but their curiosity satisfied they lay down the oars and without any farewell they let the current take them back.

This kind of sailing is in fact quite demoralizing for people accustomed to challenging conditions. A boat trip propelled by a motor along a peaceful river in beautiful weather while steering in the shade of a beach umbrella and sipping a cool drink ... that's not serious sailing. We have with us, after all, consul Kwapiszewski, who had served a long time in the consulate in Norway where he became an accomplished and enthusiastic sailor. So we leave him under the umbrella, manning the tiller while we stretch on either side of the deck and warm up in the sun.

Our laziness is soon to be punished as a dark cloud appears on our port side. Thunder and a little rain, we're thinking, as we take our pillows and blankets down into the cabin, but we are not taking it very seriously. Unfortunately, things do get more serious; the cloud grows bigger before our very eyes. A sinister silence descends all around us; the only sound the putt, putt, putting of the motor. It doesn't look good but there is nowhere to take shelter. Suddenly we are hit with a squall. Although we are not using sails the wind gets enough resistance just from the mast and the rigging to turn us completely on our beams. Moments later there are short, angry metre-high waves. We head against the wind to the other side of the river where there shouldn't be any waves. The rain hits our faces; we can't see the banks of the river. There is not a single point of orientation. Somewhere here in the middle there are shallows that we had passed earlier. But did the squall take us back to them? Suddenly the silhouette of a tugboat emerges from the rain, just for a few seconds. Very skilfully he had pulled four huge barges against its sides to protect them from crashing onto the land. It resembles a mother hen protecting her chicks under her wings.

Our helmsman, with his glasses dripping water and the beach umbrella collapsed over his head can't see anything, so he is begging to be relieved. In any case, there's nothing to see in these conditions anyway. According to the map we have no choice but to head against the wind. He changes his clothes; we give him some oilskins to put on.

Half an hour later, it's all over. The water is murky and full of leaves and branches but if not for that, one would never know that just moments ago we were caught in a very strong squall.

At sundown the situation gets more difficult. The river narrows, the water channel for ships is demarcated with illuminated buoys and is winding every which way. Our little yacht is not very big but its gross tonnage is 4.5 tons, an almost unmanageable inertia in relation to its measurements. If we were to come upon a large submerged rock it would be catastrophic. We decide that we need two helmsmen, one to steer, the other to hold the lamp and keep his eyes on the chart.

Around midnight, the buoys indicated on the chart and the illuminated buoys on the river are not matching. We've been moving for 15–20 minutes orienting ourselves pretty much by the lights on shore but after all this time we realize that we have no idea where we are.

We don't want to get ourselves into a mess so we let go anchor. The current pulls us back maybe 40-50 metres but our anchor holds; our position is solid. We take down our red and yellow navigation light and, following regulations, we hoist our white lantern high at the top of the mast, and go to sleep.

In fact, we don't get much sleep. At first light of day, around five in the morning, we immediately see what our position is. We have to start the motor but this capricious engine rejects all our persuasions. Finally it starts but after a couple of minutes instead of the normal hum it makes a racket that would drown out a cavalry charge on the stone surface in front of the Royal Palace.

We cut the engine. Everything is silent as a tomb. We can hear birds chirping. Smoke billows out of the interior of the yacht. We looked at one another. Silently, tragedy etched on his face, Świechowski lowers himself down to the engine. On deck we can hear him knocking around with something metal.

A classic scene from a film comedy: he emerges whistling as though nothing happened and places at our feet a piece of the cylinder, a piston, and half of a crankshaft.

We sit down side-by-side and warm up in the sun. In silence. We still have about 2 thousand kilometres to go and our engine is fit for a junk pile. Must we quit and drift back to New York with the current?

Five minutes later we come to a decision. We will hoist our foresail and go with the current to the nearest town—Coxsackie, NY. Consul Kwapiszewski leaves us there while he heads for Albany to get us some help. Once he gets back to New York he will demand that the manufacturer either replace the engine or make sure it is completely repaired.

Whatever the outcome, we intend to get *Dal* to Chicago, come what may. That decision has been made.

Meanwhile, we stretch out again on our blankets and go to sleep.

We continue this way for the next two days, interrupting our pleasant relaxation on our blankets only to go to the nearest restaurant for lunch and dinner. Yesterday we had guests, a lawyer, a Mr. Ogonowski with his wife and children. On hearing about our arrival in Albany, he got in his car and came down to see us. We've also had visits from a couple of reporters and photographers. All of them have been scooped by a young lady from the area.

Eighteen years old, graceful, her dress modest but chic, she arrived in a pick-up truck, "because Daddy has the car," she explained with a smile. Quickly and smartly she wrote a little article about our predicament, jumped into her vehicle, waved to us, and that's all we saw of her. She will call a few editors, and make a few dollars.

Lying on our blankets, staring up at the sky through the branches, we see a blue sky with an occasional white cloud floating by. Looking down we see the river in front of us. It is almost the same blue as the sky except for the current, which is a bit darker. Here and there the water glistens in the sun.

A ferry crosses the river, back and forth, always with two or three cars on it. Each time the helmsman waves as he passes us then turns and starts talking to his passengers. Regular customers, no doubt. One has to make them enjoy the trip, to keep them coming back. He runs his business well.

A motorboat is heading in our direction and pulls up alongside *Dal*. They have come to get gas. The name of the boat is painted on the bow, "Bonheur New York." There's a two-man crew plus a young girl. I figure now is our chance. We engage them in a conversation, tell them our dramatic history, the broken motor, and how much we are hoping for a tow to Albany. But my rather transparent story has no effect on them.

They get their gas, eat their lunch and... maybe it was the lunch that made them take pity on us because they return. The older man, the owner of the boat, slaps me on the back and says, "You see, sir, I just went on this little river trip to please my daughter and son-in-law but they decided they want to tow you to Albany, in which case this is what we must do. What do you say?" he asks. "Can we start right away?"

We don't keep them waiting. We hand our towline and cast off.

This chance acquaintance enables us to get to Albany, 50 kilometres away. We get there that evening.

Endnote

1. Elbert E. Farman, American military attaché and observer during the Polish-Soviet War 1919-20. He married André's sister, Anna, in 1951.

13

Albany–Buffalo

We are now at the Albany Yacht Club where we receive a very warm welcome. The evening newspapers feature articles about us complete with our photographs.

The city itself is an important centre, although it strikes us as strange that it is Albany that is the capital of the State of New York, and not New York City. The Governor is not in the city so we are not able to meet him but when he heard about our situation he immediately ordered that a motorboat owned by the state be put at our disposal to tow us the entire length of the canal to Buffalo. As for our engine, I am informed by telephone by Consul Kwapiszewski that the manufacturer asked to have it returned to the factory where it will be completely repaired at no cost to us.

The Polish community in Albany is quite small but a delegation from the neighbouring town of Schenectady, which lies directly on the Erie Canal, arrives to ask us to stop at least for one day to visit the Polish community there.

We sleep on the yacht that night and immediately upon awakening get to work. We will be going through more than twenty locks along the Erie and under several low bridges. We must lower the mast, heavy work that takes up most of the morning. At three in the afternoon we have two members of the delegation from Schenectady, Mr. Gwiazdowski and Mr. Mak, on deck, and we move forward; the rest of the delegation, the Ogonowski brothers, Mr. Passak and others, return to Schenectady in their cars. At this point, we are towed by a rowboat with a motor attached, its crew consisting of two policemen from the river patrol. Since this boat is extremely light and inclined to be tippy, we have to steer *Dal* very carefully. Just going even a little outside the designated channel can cause the towboat to lean to one side and get it off course.

As I get to know more about our guests I am amazed to discover that Mr. Gwiazdowski was once, as a young man, the office secretary on the Jedwabne estate that belongs to my uncle Skarżyński. He shows a great interest in my cousins: "And what is young Mr. Kazio[7] up to these days, and the young lady? I have been here now for 30 years," he continues, "but I remember everything about Poland so clearly." As he says this, he nods his head slowly and stares a long time out at the water.

We hear these sentiments expressed very often by the older generation of Polish emigrants.

We have just arrived in... Troy. Or so we are told both by our passengers and our maps. But who thought of giving this name to this shabby collection of factory chimneys and nondescript dwellings?

I am disappointed though I do realize that it's a bit much to demand that just because of this name there should be an enormous wooden horse at the river's edge, its bridle held by the beautiful Helen.

It's time for me to get used to these strange names, for example Indian names such as Schenectady, Wyandotte, Chicago (which translates as a "muddy road"), and French names, such as Erie and Detroit, their pronunciation now anglicised.

There are also Polish names: Warsaw appears several times on the map of the United States, Kościuszko, Pulaski, and there is even a town called... Panna Maria (Lady Mary).

We leave the Hudson River and enter the Erie Canal and we are about to be introduced to the pleasure of passing through the locks.

The first time the gigantic doors close behind us and we find ourselves at the bottom of an enormously deep well, our hearts almost stop beating as we watch the walls seeping water while in the middle, the water boils and bubbles.

These locks were built for large ships and for them this would be nothing. But for our little yacht this rushing water tosses our boat in all directions, threatening us with dire consequences.

Fortunately everything turns out without major mishaps. The physical effort, however, is strenuous. We constantly have to push away from the walls or pull ourselves toward the walls, constantly shortening our lines, then releasing them.

After our third lock, with each one consisting of several compartments, our gallant police escort leaves us. From this point on another motorboat, compliments of New York State, is to pull us all the way to Schenectady. Unfortunately, no motorboat is there. That is, it had been there but since we weren't... it left.

Once again we are abandoned to the mercy of God. Or so it seems. We joke about pulling *Dal* along the full length of the canal with a rope. But then, an order given by the Governor is not a joke: a couple of hours later, just as it is getting dark, a huge tugboat pulls up, the "Governor Cleveland." Since there was no smaller tug available they put this enormous ship at our disposal, a tug that could easily tow a hundred little yachts like ours.

We are truly honoured, though we are soon to discover it is really too much of an honour.

As soon as the towline is tossed to us the ship takes off at 40 km/hour raising giant walls of water all round us and under the stress of these waves our poor little boat begins to tremble in a most distressing way. We get the ship to stop. So then they attach *Dal* alongside the *Governor Cleveland*. Once again it takes off full speed ahead. Our pleas to the captain of the ship are ignored. The stubborn river wolf keeps insisting that everything is "Okay" and there is nothing to worry about. Only when I tell him that if he loves speeding so much we will cut the line and let him keep going at this breakneck pace by himself does he agree to slow down.

Finally, at three in the morning, we moor in the marina at Schenectady.

The large crowds that had gathered to greet us expected us much earlier but by midnight, they dispersed. Still, around 100 loyal Poles stayed on. A watchman is assigned to look after our boat while we are taken to the best hotel in town.

I can't begin to describe in detail every reception we received in every town we visited except to say that everywhere we received the warmest of welcomes and generous hospitality. Nor is it possible to list all the names, places and times of every reception that awaited us so I will summarize...

The Polish community in Schenectady, numbering around 14,000, started streaming in to the marina just to have a look at our yacht. These pilgrimages continued throughout the second day. The owner of a garage, a Pole, took on the job of removing our disabled engine and sending it off to the factory for repairs. When that's done, it will be delivered back to us a week later directly to Buffalo.

The welcoming committee takes excellent care of us: we are received by the Mayor of the town, there is a reception at the Polish Community Centre, and another with the Polish Sea Cadets who in turn give us contacts with the Polish Scouting organizations in the area.

From this point on our voyage demands two separate roles. On the one hand we are to continue being sailors carrying on with sailing and crossing through countless lock gates; while on the other hand, upon arrival in some town instead of resting we have to don our white trousers, our yachtsmen's cap, and our entire innate dignity and listen attentively to speeches, to give speeches ourselves, and to exchange toasts with our hosts.

From Schenectady we go to Amsterdam, towed by a man who volunteered to come up and tow us in his own motorboat. We arrive there at noon. A huge crowd, with the mayor in the lead, is gathered on shore to meet us.

A few words of greeting, some photographs, and we are off to lunch with a distinguished commissioner of the N.R.A.[8], Mr. Wytrwał, commonly known as Amsterdam's "Polish Mayor."

We visit the Polish areas, including the Polish Community Centre. A great party celebration is organized in our honour in a public park. A few speeches. The ball that evening opens with a Polonaise and I'm asked to lead. This is followed by [Polish folk dances] the oberek, kujawiak, polkas... I have a feeling that in another ten generations the Polish language here will be forgotten, but these dances will carry on.

Eventually, utterly exhausted—because one can't, after all, refuse any dance—we are led to a table at which are seated some very serious mothers and fathers. We must, of course, have a drink with every one of them.

On our way to the Golf Club where we are to have supper, I insist on driving but Świechowski gets out of the car and only gets back in when I promise that I will not drive.

When we finally bid everyone farewell around midnight and embark on the next leg of our journey, I guess I don't have to tell you that we have a hard time handling the yacht.

In the morning we arrive at the biggest locks in the United States. When the water in the lock finally reaches the top, without a mishap, I can honestly say that a big weight comes off our hearts.

That evening we are in Utica. As in Amsterdam, a large crowd awaits us at the marina. A rollicking dinner at the Polish Army Veterans' Club. We go off to the park to have a look at the monument to Pułaski, and then a reception at the Polish Community Centre.

We depart from Utica at 8:30. Now we are towed by a tugboat provided by the State of New York. We are surrounded by a magnificent, forested countryside. Along the shore are lovely wooden houses—summer cottages. Having read about our arrival and passage in the morning papers, we are greeted by people on shore who are calling out to us and waving.

In the afternoon we sail across Oneida Lake. For us, Europeans, such a lake seems immense, about 35 kilometres in length. Its majesty is dwarfed, however, by nearby Lake Ontario, which is ten times bigger.

There's a heat wave. We let go anchor a distance from shore. Our tugboat captain goes into the nearest town to report to the authorities on our progress and that everything is in order. We take advantage of this stop to get out of our clothes and dive into the water.

Once underwater, we see that the copper siding on the boat is coming loose in some places. To ensure this metal sheet does not hook on anything while we are sailing we have to dive in with pliers and break off these pieces.

As expected, all the seaweed and shells that attached themselves on the voyage from Bermuda to New York have fallen away in fresh water. When we have finished these tasks we get back on deck and warm ourselves in the sun, waiting for the return of our sailing comrades.

Once we exit the lake we moor at Breverton, right by the bridge. Only a few Polish families live in Breverton but the entire cream of Polish society in Syracuse drove down in their cars to greet us.

We are invited for a visit on the deck of a motor yacht owned by a Dr. Krzywicki, who sailed down from Syracuse to meet us.

This is our first time on a sporting boat owned by a Polish American. In the evening we go off all together to the Hunter Club for supper. After a few obligatory speeches to officially mark the occasion, the evening is very relaxed and we all have a grand time.

As soon as we finish supper, we pile into the car and head off to Syracuse, some 30 km away. We arrive at the Polish Community Centre where all those who were unable to go to Breverton are gathered to give us a rousing welcome.

As we leave the "Polish House" we are told that we are going to someone's country house. With Polish hospitality in the traditional Polish style, our host, Mr. Gimiński, is putting us up for the night. In the spirit of it all he wants this to be fun, with dancing, so at the last minute he invites several guests and improvises a real party.

We look around the farm. Were it not for the splendid bathroom decorated with blue tiles, and a row of cars parked in front, one could think we are at a manor house in the Polish countryside.

We have a marvellous time, dancing until late into the night. Since our host is a bachelor, everyone helps out as much as they can, some making sandwiches, others pouring drinks.

We remove our jackets, American style, and drink large glasses of whiskey and soda on the veranda. We all feel so good together that before going to bed we persuade Mr. Gimiński to join us next day all the way to Rochester.

In the morning we are awakened by reporters from the local papers. "Hello, hello! How are you?" and they sit down on our beds. They tell us that Świechowski looks a bit pale; as for me I find it difficult to explain to them that I begin every morning drinking several glasses of water with some baking soda. Finally the words, “training diet” registers with one of them who, nodding his head, diligently write down what we ate for our first breakfast.

When we arrive at the dock we realize how foolish it was, on our part, to invite our host to join our voyage to Rochester, and how foolish it was, on his part, to accept our invitation.

Our new passenger, you see, weighs about 130 kilos, but it is too late to back out. The motorboat provided by NY State to tow us is already waiting for us. Our guest jumps into our yacht with an unexpected agility, considering his size, so we quickly change into our working clothes and cast off on the next leg of our journey.

There is a lot of river between us and our next stop, Rochester. We'll have to go non-stop if we want to get there by nightfall of the next day.

When we left yesterday for the farm I couldn't find my toothbrush. Taking advantage of Świechowski's turn at the helm, I search the cabin, turning it inside out. No easy task because the cabin is an absolute mess. My search is like the proverbial search for a needle in a haystack. Then I remember that the previous day, when we were swimming in Lake Oneida, I had placed it in a hiding spot on deck next to the mast. Imagine my surprise when I find in this place a small, carefully wrapped package containing two brand new toothbrushes with a card: “I took the old one for a souvenir.”

The day is unusually hot. For Świechowski and me, this is just our daily bread, we accept conditions as they come, like a couple of old philosophers. But our guest does not take this philosophically. From the start he curses the world in which he lives, can't stand heat, and wants "something" to drink. With genuine naiveté and a kind smile Świechowski offers him a glass of rather warm water from one of our containers, which is rejected with scorn and coarse language.

Unfortunately, in the flood of words colouring his displeasure, we make out one little word: "Beer."

He does not realize that saying this was like handing us a weapon of torture, that suicide would be preferable to the suffering that he was about to endure.

So far as I recall I have never had any talent for poetry but that day I am truly inspired. I don't know why. I begin emoting about the ecstasy of being in a cool, quiet American bar with a litre jug of beer in hand, with its snowy white foam on top and misty dew on the outside of the glass... words that would have enchanted many a beer lover. Świechowski loyally echoes me.

At first our companion stays calm. But his expression, as he glares at the cabin, is so horror-struck we think he is seeing ghastly ghosts.

At that point he starts a terrifying roar in his deep basso voice, "Please, in the name of God, please stop..." so much so that it resonates throughout the surrounding forests and even the crew on our towboat, despite its roaring engine, is signalling signs to us trying to figure out what the problem is.

Our problem is finally resolved that evening at a pleasant restaurant on shore where it is demonstrated to us that one can drink several litres of beer before eating.

After our supper, we move on but have to think about our overnight stop. Speaking for ourselves, we could easily keep going with the usual watches, one at the helm, the other sleeping in a bunk, but our friend cannot squeeze through the hatch to get into the cabin. He has to move to the motorboat where we arrange a nice place for him to sleep.

We finally arrive in Rochester the next day at four o'clock where we moor at a large marina next to a very large hangar. A large crowd awaits our arrival and gives us a royal welcome.

The committee greeting us is made up of representatives of every local Polish organization, of which there are a great many in Rochester, as well as representatives of the Rochester Yacht Club with Commodore W.P. Barrows at the head to give us a very heartfelt welcome. The celebrations are interrupted suddenly by an unexpected rain. For us, dressed in our working denims, this is not a problem but the elegantly attired ladies are most discomfited by it all. Fortunately almost all of them have their own cars so within minutes all that remains in front of the hangar is a long row of automobiles, their occupants honking, shouting and waving at us.

We take advantage of some free time to secure our moorings and carry our wet sails into the hangar where we hang them out to dry. We also take out our dress clothes and change in the office of the hangar.

While we are changing the hangar fills with reporters and photographers, as well as masses of people who begin thinking of us as their "unknown close friends." We have never seen them before, nobody knows who they are, but despite that they are so friendly and familiar that we are inclined to ask them about their health, their wives, their children.

The photographers are allowed entry only on the condition that they refrain from taking photographs until we look decent, so while they wait we are in the hands of the reporters. And there were a lot of them! Rochester and Buffalo are, after all, the biggest cities in New York State, after New York City itself of course. There were reporters from both these cities, among them Mr. Ruszkiewicz, the owner and editor of Dziennik dla wsystkich, (Everybody's Journal), one of the largest Polish-language newspapers in the United States.

When the rain lets up a little, we get into a car and are driven to the Stevens Hotel where we find everyone at a great banquet.

The mood is upbeat, and very festive; all these distinguished gentlemen had left their homes in Buffalo or other nearby towns and are very much in the mood to enjoy themselves and, one must say they achieve their objectives one hundred per cent.

We begin with a wonderful Polish-American celebration after which we go to our hotel and are shown to our rooms. Well, they are not "rooms" but luxurious, adjoining apartments.

In the middle of the main rooms, in both apartments, there is an enormous canopied bed. On one side is a door to a green-tiled bathroom, on the other a door to a boudoir and then to a salon. There's enough space for six couples to dance the mazurka.

We are escorted to our hotel by a large pack of young people so we decide that we must have a party. We get a couple of bottles of excellent Scotch whiskey from our stock on the *Dal*—a gift from our friends in Bermuda and likely not available here. Every respectable American hotel has a radio in every room so we turn it on loudly.

Around midnight, just as the party is in full swing, our telephone rings: "This is the manager of the hotel," we hear a deep masculine voice say, "Are you gentlemen aware that according to local regulations keeping women in one's room after midnight could result in a severe penalty?"

I am terror-stricken. I remembered all the stories about American laws about which we had been warned in Bermuda. They will throw me in jail, Świechowski will be ordered to marry every one of these young ladies and adopt every one of the young men, and also compensate the parents of the whole bunch for so suddenly and terribly taking away their children and then pay another compensation to all the fiancés for breaking their hearts. My forehead was suddenly dripping with sweat ... need I add that within moments we are completely alone?

There's a knock on the door. Before I have time to say, "Enter!!" the door opens and, nonchalantly, in walks our good friend, the captain of the local police, Mr. Felerski. The sleeves on his shirt, blue-striped with white dots, are rolled up. He is chewing on a very big Havana cigar which he is moving, as though with distaste, from one side of his mouth to the other... We fall upon him as though he were our last and only hope. He listens attentively and patiently to our complicated explanations occasionally nodding his head with all the dignity of his high office and when we finally have no further arguments to make in our defence he removes his cigar from his mouth and says:

"This kind of situation is indeed very serious but in this case there is something very much in your favour, and that is that I am the one who made that call. You see, it was quite past bedtime for those youngsters while in the next apartment there is the entire old guard waiting for you. Come, you will see, and you won't regret it."

And indeed we did not.

When we take our leave from Rochester next day the weather is marvellous. Once again there is a crowd at the marina gathered to bid us farewell.

The delegation is headed this time not by Mr. Gimiński but in his place is Mr. Ruszkiewicz, the publisher of *Dziennik dla Wszystkich* [Everybody's Journal]. We are sorry to say goodbye to our former passenger whose good humour is not at all affected by the previous day's difficulties. A reporter from one of the American newspapers from Buffalo joins our expedition but he is in the motorboat that is towing us. He is to keep his editor abreast of all our doings.

We don't yet fully understand the power of the popular press. In Poland, the press does not cater very much to the public as it does abroad, especially in America.

Here, although a journalist risks serious consequences in his work, even the lowliest journalist has access to people of any rank in the United States. But in Poland, a journalist once told me, "it is not worth trying to get an interview. They keep you sitting too long in the waiting room."

The canal is now passing through areas heavily settled by Polish immigrants. All of them have already heard about our arrival from their newspapers. We, and Mr. Ruszkiewicz who is with us, are very surprised to hear greetings shouted to us in their mother tongue. "Good day! Long life! God be with you!" and from the younger generation, "Hello guys! *Jak sié macie?*" (How are you?)

In an open field just before Albion, there is a priest surrounded by his parishioners; it seems they all ran together from the nearby church just to see us. His eyes are full of questions: "How is it going in Poland?" is all he manages to say and without waiting for an answer he gives us a sad smile and makes a sign of the cross to bless us on our way. He will remain in my memory.

After dinner in Albion, when we get back to our yacht we are presented with a very large torte with ice cream, and despite our best efforts and that of the crew, we are unable to finish it.

After Albion, there is a road parallel to the canal on which there is a slow moving caravan of cars with people waving to us. After dark, they occasionally flash a spotlight in our direction. Around midnight, while we are struggling with foaming water in the lock, the darkness is sharply pierced several times by the flashbulbs of journalists who will soon rush into their cars to get their stories off to their editors in time for the morning papers. One of these reporters even comes down along the slippery wall of the sluice grasping the metal hooks and, uninvited, jumps onto our deck and starts helping us. Son of a gun, he handles oars as gracefully as he does literary paradox. During the time the lock was filling with water he manages to get enough material out of us to file his story for the next day's newspaper.

To avoid arriving in Buffalo during the night, we stop in Lockport. In the morning we push forward. It appears that there is a royal welcome awaiting us. When we are out of the canal, we sail past some Polish fishermen pulling out their nets. In the distance, on the right, we see the Canadian shore, on the left is the marina. A large motorboat comes out to meet us. Mr. Ruszkiewicz identifies all the important people from the city. There is Mr. Zimmerman, the mayor, along with several Councillors, senior government officials, and representatives of the local Polish community. We exchange greetings, then the motorboat goes in front of our tugboat and escorts us into the marina.

This is quite an honour. Buffalo is, after all, one of the richest and most beautiful cities in the United States. Thirty-seven percent of the population is Polish.

Arriving in Buffalo

We sail past American Naval Cadets in dress whites standing at attention in front of their mast. They greet us with a gun salute and by lowering the American flag; unfortunately our mast had been removed so we are unable to signal our greeting.

Finally, we sail into the Buffalo Yacht Club marina. There are several thousand people shouting their greetings, along with the deafening whistles and sirens of nearby boats and ships. We moor our yacht, we meet the mayor, the various state and municipal officials, and the representatives of the Polish community. We are also delighted to shake hands once again with our old friend, the Commodore of the Yacht

Club, Dr. Hussey, with whom we spent several enjoyable months in Bermuda.

Polish organizations are out in full force, including the newly established Polish Sea Scouts as well as the young people of the Maritime League; their girls in beautiful uniforms present us with a great bouquet of roses. There are also representatives from Italian and German groups. The latter hold our hands tightly and tell us they came especially to stress their support for mutual friendship between them and the Polish community. The Polish colours are hoisted on the flagpole of the Buffalo Yacht Club where it will remain throughout our visit to the city.

Endnote

7. Kazimierz Skarżynski, later wartime president of Polish Red Cross, wrote the first official report on the Katyń massacre.
8. The N.R.A. here is President Roosevelt's National Recovery Act, and the commissioner is Michael J. Wytrwał.
https://en.wikipedia.org/wiki/Michael_J._Wytrwal accessed Feb. 17, 2019.

14

Buffalo–Detroit

From Buffalo all the way to Chicago our route will be entirely along the Great Lakes. No more low hanging bridges to go under so on the day we leave we step our mast once again and hoist on it—as required by protocol—the American colours, and our Polish flag over the stern.

A couple of days after our arrival [in Buffalo] we read a story in the newspapers about our flag that is at once stirring and terribly funny, though in fact, it could have ended tragically. So... after a mighty bout of drinking two old Polish men decide to go have a look at *Dal*. When their eyes behold the dear old red-and-white flag of their distant native land, their patriotic feelings are so aroused that they feel compelled to kiss it. Unfortunately, just as they get there the poor old guys lose their balance and fall into the water, still clutching our flag. It all ends with a good night's sleep at the local police station for them but our flag is only rescued after a lot of diving and searching by some young members of the Maritime League.

Our stay in Buffalo lasts two weeks. The manufacturer of our motor, the Sterling Company, returned it to us in good working order and it is re-installed thanks to the local Polish community. In the meantime our stay is filled with sightseeing, celebrations of all sorts, and speeches. We get up every morning between 8 and 9. While we bathe and dress, several gentlemen wait to take us back, and a few reporters ask questions: "Hello boys, did you sleep well? Where are we going today?" At this point our patron, Mr. Kaszubowski, interjects, "Now boys, let's have something new. Something that hasn't appeared in the papers yet."

This is a frightening moment because we must come up with something quickly otherwise he'll come up with something about our voyage in great detail none of it making any sense. We can't risk opening a paper some day and reading that we fought off an attack by flying fish, or something like that.

The days fly by. We drive to Canada to have a look at Niagara Falls. We are accustomed to borders when travelling. Few people in Warsaw realize that the capital is only about two hours away from the beautiful Mazurian Lake District and yet practically no one has ever been there. We expect a few problems here too because we have no passports, only the documents for our yacht, *Dal*. As it turns out, we don't even have to show them. By a weird coincidence the customs officials heard our speech at the Yacht Club on the radio. So a pat on the back and handshakes all around and the border formalities are taken care of.

Later that day when we are describing the impression Niagara made on us one of the fellows proposes that we take a flight over the falls. We don't need to hear this twice! Before the day is out, the plans are in place.

On the outskirts of town there is a gas station owned by a Mr. Kucio who lives in a little house beside it with his wife, his mother, children and brothers. If not for the paved highway and the steady stream of speeding cars on it, the scene is reminiscent of a sleepy Polish village. Behind but not far from the little house is a barn in which there is an aeroplane. Pulling all together, we get it out. A moment later there is a deafening noise from the engine and we are up in the air, so high that the gas station, the house and the barn are but little toys beside the long ribbon of the road. Next we are flying over the city and a moment later we are above the falls. Strong winds are tossing us around mercilessly but the view of the mass of thundering water from on high is simply breathtaking; our little plane is circling a couple of hundred metres above it.

A few days later we visit a local sight of a very different nature, the prison in Attica. Never in my life have I seen such comforts and were it not for the rows of cells and the thick clubs carried by the guards I would think we are in a first rate hotel. The variety of workshops and classrooms amaze us. Had we not noticed the occasional furtive glances we could easily forget that we are in a correctional institution.

But the ultimate surprise is the prison hospital. Świechowski is so impressed that he eagerly inquires what kind of a crime one has to commit to get into this kind of prison, what kind of illness one has to have to get access to an ambulance.

Just as we are walking past an operating room a fellow with Semitic features wearing a white apron comes bounding out. He throws his arms around us, warmly shakes our hands. “Gentlemen! My fellow Poles!” he says in the purest, well educated Polish, gesticulating enthusiastically. “We are so proud of you; if only there were more like you!" We thank him wholeheartedly for his greetings and as we go downstairs I discreetly ask, "Who was that gentleman? A doctor, I suppose?"

"No," replies our guide, "Not a doctor. A banker." Thinking for a moment he adds, "He has another eight years. For passing bad checks."

Sic transit gloria mundi.

One day our friend from Bermuda, Dr. Hussey, takes us to The Canoe Club. Despite the name, it is not a kayaking club but a yacht club. The membership of the club is exclusively American but because the Canadian shore of Lake Erie is nicer, they built the club in Canada.

We sail over in a yacht belonging to the doctor's grandchildren, who are now adults. They promised to return our visit some day in Gdynia. "You will?" said Dr. Hussey with a doubtful smile. "Well, in that case all I need is to build myself a new yacht and I will go too." He is almost 80 years old now but considering his recent feats he quite convinced us that if he just wanted to, he could.

As the hour of our departure approaches we make a series of farewell visits then cast off on the long voyage ahead. What a joy it is to sail once again with wind in our sails rather than with the noise of a motor as we enter the vast, seemingly limitless, expanse of the Great Lakes.

At first glance one could think that having sailed across an ocean, the lakes should pose no difficulties: that would be a serious mistake. The Great Lakes in America are actually fresh water seas with sudden intense storms. Because the lakes are relatively shallow, the storms are unleashed not over a period of several hours, as on the ocean, but in a matter of ten minutes. The waves they create are short but steep so that smaller boats can't settle across on them. Woe to the yachtsman who is caught in a storm with full sails.

After October 1, insurance companies refuse to insure even large steamships because from that date and right up to May storms ceaselessly erupt across the lakes. Which does not mean that there is a shortage of them during the summer months.

Our next stop is in the small town of Dunkirk. We arrive in the early evening accompanied by Mr. Kaszubowski, the lawyer, and Mr. Janicki, the president of the Maritime League in Buffalo.

We receive a heart-warming welcome, greeted by several town officials who had arrived in their cars. They take us in a parade with the town orchestra marching ahead of us all the way to our hotel where there is a splendid banquet arranged for us. We spend the night there and early next morning sail off once again.

The weather continues to be favourable. This time the guests on deck are two Polish Americans from Dunkirk. One of them Judge Johnson, is a descendant of one of the first Polish families to settle here but who, as unfortunately is frequently the case, had changed his surname to fit in more easily among the suspicious Anglo-Saxon colonists and be able to obtain better positions. At the time of his graduation from university, Mr. Johnson didn't speak a word of Polish but subsequent events in his life reveal that one never knows when the feeling of belonging to a nation will surface again. He married a Polish woman and now speaks Polish, as do his children, and was elected to his position thanks to a large Polish vote.

We arrive at our next port-of-call, Erie, at twilight. Several large yachts sail out to us, greet us in the traditional style, and guide us to a place in the marina reserved for us.

As soon as we made fast our boat and step on land a huge crowd surrounds us, pressing from all sides. To facilitate our passage, we are guided against the crowd toward the steps of a harbour crane. From this improvised dais, facing the blinding lights of flashbulbs, we utter our words of greeting.

A large hotel room had been reserved for us where ten gentlemen who accompanied us promptly sit down on every chair and bed. A moment later a waiter enters bringing us whiskey and soda. Before long everyone is noisy and cheerful. We are, by now, accustomed to this sort of thing. It doesn't even interfere with our toilette.

Through the partially opened bathroom door I can hear Świechowski splashing like a whale but I take advantage of this time to shave and to send off our clothes for laundering. Then just as I had taken his place in the shower I hear him wailing in despair and horror. By mistake I had sent off his white trousers.

My grievous error is soon corrected. Within half an hour we are able to get him a new pair of white trousers from a store in town. We get to the Polish Community Centre a bit late because of this but this also gave us something to joke and laugh about. Next day a headline in a local paper reads: The Hero Lost his Pants."

None of this would had been very important to me except that for some reason people identified the loss of his trousers with me. The next day, after our visit to the mayor's office, accompanied by the Commodore of the Yacht Club, as we cross from the hotel to a restaurant we pass a group of young American girls. One of them shouts to me: "Where did you find your trousers?"

What happiness!

Next day we set sail for Cleveland. This time by ourselves, giving us some free time to restore order to our boat. Strong wind. To get more speed we get the motor going and that, together with the sails, has us going at 20 km per hour. We speed by the white band of sand dunes along the coast quickly. Checking our map we figure that if we keep up this pace we will arrive in Cleveland during the night, but we are expected to arrive only at noon.

We decide, therefore, to stop at the little Firestone harbour, on calm water deep inside the canal. There, among barges and tugboats we spend a restful night.

The bangs and clangs of the motors of several fishing boats and the rocking of the water abruptly awaken us. The fishermen's flotilla is just heading out to the lake, so we follow shortly after.

The wind calms down. Sailing at a relaxing tempo we enter the port in Cleveland at the arranged time.

Once again we enjoy the same enthusiastic greeting we had received in all the bigger cities with large Polish communities.

There are two Polish-language newspapers published in Cleveland. Some sections of the city are so densely populated by Poles that they are known by their Polish nicknames. Hence, people say, "Let's go to Warsaw," or to Krakow.

We spend a very pleasant evening in the home of some veterans of Polish armies. Among them I am delighted to come upon some veterans of General Haller's[9] army who returned to America after the war.

Our visits to the mayor of the city, several of the larger Polish parishes, and to the large convent of the Josephine Sisters who ran a school for girls very professionally, absorb pretty much all our time in the city of Cleveland.

The highlight of our visit is the official reception organized by the League of Polish Organizations in the theatre of the Polish Community Centre, attended by thousands of people. Despite having people stand in the aisles and at the back, there is not enough room for everyone who came.

After a few speeches of welcome by the president of the League, a Mr. Rella, the Editor, Mr. Przywary and others, the deputy mayor of Cleveland, Director Kurdziel, presents us with a gold key to the city.

After four days we set off again on our voyage, this time heading north across Lake Erie to the mouth of the Detroit River and then up to the city of the same name. We are not going directly there but stopping along the way in Wyandotte, a town a little distance ahead of Detroit. We intend to stop for a couple of days rest there with some friends. We have 150 km ahead of us so leaving in the afternoon we'll be sailing through the night.

That evening we are hit by a powerful squall that lasts about half an hour.

That gives us our first introduction to the storms of the Great Lakes. In a matter of minutes the gently rolling surface waves become violent. The waves are so short and steep that *Dal* cuts into them deeply. We leap to reduce our sail as much and as quickly as possible, attach the canvas covers as needed... and almost as soon as we get everything done, the wind dies down, the waves are gentle again. We are soaked to the skin, and all for nothing.

At nightfall, the shore has long disappeared from view, the promised lights are nowhere in sight so I steer very carefully following the compass. Surely we'll see them soon.

Świechowski wants to sleep so heads down to the cabin without waiting to see the lights. Anyway, he will take over the watch at midnight.

The time drags on, one hour after another, and still there are no lights in sight. Straight ahead of me on the horizon, there are long, low and dark clouds forming, gradually getting bigger. The top of the cloud is strangely jagged. My eyes are trying to pierce the darkness enveloping us and then, suddenly, I understand. That is the shore. What I took to be a jagged cloud is a shore lined with trees.

I turn sharply, the boat obeys my command. I call Świechowski.

We curse the world and all it stands for. What a disgrace! We have sailed across half the globe and then we get lost in a miserable pool. As we later discovered, the compass wasn't functioning properly. We had placed it next to the engine and from that point on we never checked it.

It didn't occur to us that navigating parallel to a shore we would need it. In the genial atmosphere of banquets and merrymaking we forgot one important thing which now is a lesson for us to remember: until the moment when we actually get to Chicago, we are first and foremost sailors.

In the morning we enter the Detroit River. We are sailing upstream against a strong current. The wind has died down; we are using only the motor. On one side we see English flags, on the other American. Here is Canada, and there is the state of Michigan.

A large motorboat is speeding towards us, shows us where we are to stop. We are to go around the island. We make a circle and go where, according to our map, there is a depth of four metres; unfortunately, we didn't know that because of a long drought the water level on the river has dropped. So we ram the keel into mud. What can you do? I take off my clothes and lower myself into the water. Finally, after a terrific struggle and with the help of a rope thrown to the motorboat, we pull ourselves out of the shallows.

Not long after that we arrive at a small pier for rowboats where we are heartily greeted by a crowd of Poles from nearby towns, some 50-60 families in all.

Those couple of days in Wyandotte finally give us the opportunity to get a real rest. The Polish family with whom we are staying does everything to ensure we can relax. It is completely peaceful. Only in the evening, after our supper, do we receive a few dignitaries—such as the baker, the butcher, the owner of the restaurant, "The White Eagle." A few close friends of our hosts, including a brother and sister team of lawyers, the Kurzatkowskis, are still our friends.

We also have a visit from a representative of Standard Oil, who was instructed by this powerful company to give us a detailed plan of places where—if the need arises during the rest of our voyage —we can get more gas for our motor.

On August 3 in the morning we continue our voyage. A motor yacht, flying the red and white Polish ensign on its prow, comes from Detroit to greet us.

We are received at the Detroit Yacht Club in grand style, following all the yachting etiquette: cannon fire, saluting with flags. In the large clubhouse a splendid banquet awaits us, with 100 people in attendance.

As we enter, a beautiful little girl dressed in a colourful Polish costume presents us with flowers. After the banquet there are several speeches, both in Polish and in English. And that ends the marvellous evening.

We then drive to a hotel. As soon as we enter, our eyes alight on the strangely outfitted guests. They all have very serious expressions and three quarters of them are wearing fezes and Turkish garments in a variety of colours.

One could be merciful and forgive the men, but the elderly ladies in their scarlet harem pants and gold-embroidered little jackets that barely pull across their imposing bosoms look a bit... well, out of the ordinary. One made an especially strong impression on us in the elevator. Her turban drooped down over her eyes and under her arm she clutched, for reasons unknown, a saxophone. When the elevator attendant asked her where she wanted off, she replied: "In seventh heaven."

We help her out on the seventh floor with all due respect given her age and, undoubtedly, her position in the Sultan's harem, then rush back into the elevator to go up to our "fourteenth heaven," the paradise for tired philosophers and sleepy sailors.

We do finally meet a few "Turks" and then the mystery is explained. It is a social club[10]—there are a great many such in the United States—and the occasion is their General Meeting. But why do they dress like Turks? I never could quite understand the explanation.

Our stay in Detroit is a bit different from our visits to other large American cities. Summer is now in full swing, a time when all the companies and organizations organize picnics in the nearby forests. It became our duty to attend all these events. Very often the most tiring part of these occasions is signing autographs for the kids. For over half an hour we have to sign scraps of paper, handkerchiefs, bags... but at times I have the honour of signing an autograph book where there are other autographs by personages such as Wilson, Paderewski, Roosevelt, and even Dempsey, Garbo, Fairbanks, etc.

But the request that pleased me most was from a Pole, a sailor in the American Navy, who asked for an autograph on his shoulder, which he would later have tattooed.

A couple of days later the Polish Consul from Pittsburgh arrives, together with his wife. I am very pleased to meet him at last having long corresponded with him asking for advice about our activities here. He accompanies us to several more community picnics and other events.

We make several official visits, and tour a Ford plant.

It's hard to express in a few words the impression made on me by these auto plants. Not infrequently one is struck at once with a sense of awe and of gloom. It's not just because it is a mass of machines and people, but it is a kind of clockwork mechanism on a frightening scale. Every movement made by a worker is studied and specified. For hours and days on end the "free" man-machine repeatedly performs with great precision four or five identical moves. It gives the impression that once a man enters the factory he hangs up not only his cap on a hook but also his soul. Well, why not? What would he do with it here?

Is the man-machine happy? Yes, it seems to me. He has a modern house, a radio, his own car. He can fill his free time with mechanical music and movies.

I wonder, though, what kind of influence will this kind of work have on future generations?

We meet a Polish transoceanic pilot, [Stanislaus] Hausner, who is making preparations for his second attempt at a solo transatlantic flight—the first, two years ago was unsuccessful. He strikes me as a calm, serious guy, not one inclined to do anything foolish. We never expected him to have such a tragic end before his projected flight. [Hausner crashed shortly after take-off.]

He asks me to join him in a radio interview, saying it would help him with publicity. I readily agree.

He arrives next day at our hotel with his manager. The interview lasts about 15 minutes, the format being a conversation between someone who just arrived from Poland and another who is soon planning to fly there.

After they drive me back to the hotel, [Hausner's] manager takes me by the shoulder and says, "Mister, what are you going to Chicago for? All kinds of people are going to make money on that, except you. I have a proposition for you: a six-week tour of music halls and theatres. All you'll have to do is talk the way you just did on the radio. We'll rent a yacht. It will be part of a travelling exhibit of motorboats touring all the bigger cities. I guarantee you $35,000."

He is amazed when I turned down his offer explaining that I had made a commitment and have a moral obligation to Poland.

"Mister, think again about my proposal. In a couple of years nobody will remember you or what you did. But money is money..."

We are now invited to Hamtramck, a little town on the outskirts of Detroit. This is probably the most Polish town in the world, its population 98 percent Polish.[11] A delegation is sent out to meet us. When we get downstairs we see that, along with the cars, there are several policemen on motorcycles. As soon as they see us they approach us with friendly smiles and give us a warm handshake. Every one of them a Pole, they came to provide us with an escort. Once we get into our cars, they line up ahead of us on their motorcycles.

"Okay?" they ask. "Okay," reply the drivers. And we're off, full speed and non-stop siren whistling. Everywhere we go the traffic is stopped, barriers are raised to let us through. Half an hour later we pull up in front of the Town Hall in Hamtramck. All the city officials are waiting for us in the Council chambers, including the mayor. It is an open session and when the words of praise are read out to us, there is unanimous approval.

In the evening there is a banquet at which I meet many interesting people, among them the representative to Congress in Washington, [George G.] Sadowski. He presents me with a little known general history that I still refer to from time to time.

As a matter of fact we know very little about American history. We have a fixed idea that America is an Anglo-Saxon country with immigrants from other countries arriving only very recently.

Polish Americans think the same way about their country's history, mentioning Kościuszko and Pułaski as two notable exceptions.

A few months later when I am addressing a large crowd in the Federal Building in Chicago on the occasion of the anniversary of General Pułaski's death I make a point of stressing that Pułaski was not an exception but in fact is emblematic of many Poles who fought for and defended the United States with their blood and sacrifice.

Very few people know that as early as 1608 about 400 Poles settled in Virginia, that along with Pułaski and Kościuszko there were about 200 other Polish officers serving in Washington's army, over and above the rank and file. These officers acquitted themselves honourably; indeed six of them were killed in battle.

During the American civil war of the North against the South, among the first to fall on the field of honour were Captain Bladowski and Tadeusz Strawinski. Among those who served with distinction were Generals Krzyżanowski, Karge, Tochman and many others.[12]

Poles, numbering 85,000, served in the Spanish American war. Among the first 100,000 volunteers in the [First] World War were 40,000 Poles.

Yes, Polish Americans are Americans, in the full sense of the word.

Because of the large crowd expected to see us off, we bring our boat from the Yacht Club and moor it in the pier of the River

Patrol. We leave the city on August 9, at 3:00 pm, accompanied by a friend's motorboat.

Endnotes

10. The Shriners, a fellowship best known for their hospitals for children.
11. Today Hamtramck is multi-ethnic with a large Muslim community and is a model of multicultural harmony. Polish American historian, Karen Majewski, was re-elected mayor in 2017.
12. All but one, Tochman, fought on the Union side. He was strongly criticized by Polish émigrés.

15

Detroit–Chicago

Sailing north, we cross Lake St. Clair, just 25 miles or so, and by evening we are entering the St. Clair River. On our left, the shore is lined with villas of an original design, built on stilts. Each one has a little harbour for motor and sailboats, many of them presently sailing beside us.

It is dark by the time we find the gas station that is to service us. Despite its being past 11:00 pm, our engine is thoroughly checked over, oil and gas attended to. While this work is going on and we are relaxing on the pier, we are approached by a few swarthy, ragged young boys. They are strangely serious and pensive. Smoking some cigarette ends, they speak quietly and gently. They know the river and the shoreline so well that we assume they are raftsmen.

"No," one of them tells me, shaking his head. "We live over there, on the other side. On the reservation. We are Indians," he adds with a sad smile, "You know, the people who, on Sundays, put on feather headdresses and dance..."

We lapsed into silence. The only sound was the lapping of water beneath the pier. From the river came a rush of cold air, smelling of a swamp. We were overcome with sadness. This was our first encounter with the original owners of this land.

There will be no more Polish settlements along our route so we decide to move on as rapidly as possible, stopping as infrequently as possible. Unfortunately, bad luck is following us. In the afternoon, while closing alongside the pier we fail to notice a long iron rod embedded in the concrete. A strong current throws us against the rod, making a crack on our topsides just above the water line. We sacrifice an hour's time to caulk the hole with oakum and cover it with a sheet of metal.

The wind is getting stronger, blowing straight into my face. Nevertheless, we are leaving our little safe harbour and heading out of the river to the next lake. Lake Huron is bigger than Erie; we have 400 kilometres ahead of us on it.

We are using only the engine because the wind and the waves are such that despite all our efforts we are not moving forward. At midnight we give up and return to our port on the river. At dawn the wind changes direction so we try again to get out and be on our way. This time we have more success and by nightfall we stop at Harbor Beach because we feel we must get some sleep.

There are some fishermen on shore forecasting the weather. Two of them are not from here. We quickly get acquainted and go out together for supper. As we sit around the table they tell us a bit about themselves. They bought their boat here and are now on their way home. They will be following the same route as we will, only not all the way to Chicago. One of them, the older one, was an aviator during the war. [WWI] After the war he came back here and transported some goods, using his own car but after prohibition ended there was no money to be made doing that so he returned to his old work.

Fishermen on the Lakes gave up sailboats a long time ago opting for engines, and very powerful ones at that, because of the unexpected and violent storms. They use this tactic: after studying the sky carefully so they can tell what to expect in the next hour or so, they leave port and throw out their nets. When they see a storm coming they speed back to port in their fast boats, doing 30-40 km/hour.

Next day we are the first to leave. The weather is splendid and an hour later we see two lads speeding past us. The younger one, a very funny guy, started running the length of the boat executing some kind of strange manoeuvres. Despite the rocking of the boat he jumps from bow to stern and back, bending over, straightening up in our direction. Not wanting to be outdone, we decide to signal our farewell not by using international code flags but by hanging an old pair of trousers from the crosstrees. As soon as he sees this he gleefully started jumping up and down with such energy that he nearly fell out of the boat until his friend gets tired of this and pulls him by the leg into the cabin.

That evening we stop at Alpena. Our friends don't join us there. We spend the night tied up to two pilings alongside some kind of coal storage. It is raining, we are gloomy and wet.

There's a weather station in Alpena that issues reports every day at noon, so we go there. The director is a German from Upper Silesia. It is a pleasant surprise to be greeted so warmly, like compatriots. While we wait for telegrams to arrive, he shows us all his instruments and asks us about Katowice[13],which he hasn't seen in forty years. He speaks to us in English though inserting from time to time some German, and even some Polish words.

After studying the weather forecasts and checking his instruments he informs us that a storm is expected, but not within the next twelve hours or so. We exchange our good-byes and set off immediately in order to reach a port this evening before the storm starts.

Alpena is set on the innermost point of a bay cutting into land by about 20 km. Before we manage to get out of it, within two hours something goes wrong with the weather.

The wind, blowing in strong gusts, makes us turn north again. It is also raining. The sky is completely overcast. We've covered 50 km so don't want to consider going back but we soon discover that getting to our next port is but a dream.

Examining our map, we see a small and shallow bay just a few kilometres ahead of us and turn in that direction. With every passing moment the wind gains force. The waves are short and steep, our bow cuts into them every minute; this slows us to such an extent that we are barely covering two kilometres per hour.

After a few hours of battle we turn into the bay where the water is totally calm. Moving very slowly, the whole time sounding the bottom, we glide in as far as we can, right to the mouth of a little stream flowing into the bay. We decide to spend the night there and let go two anchors. The situation turns out to be more complicated than it seemed at first sight.

So long as the wind keeps blowing from the north, with the peninsula sheltering us, we can rest easy. But should the wind suddenly turn and blow from the east—as often happens on the Great Lakes—in a matter of minutes our peaceful bay will turn into a cauldron of bubbling trouble in which case we'll have to escape from here as quickly as possible and battle the storm on open water. In the meantime, we have no choice but to prepare for battle and listen to the ominous howling of the wind as it blows across the peninsula and into the top of our rigging.

We sleep in turns, checking from time to time to see if the wind is changing directions. In the morning, exhausted, we sleep like rocks. By the time I wake up, it is complete daylight. I leap from my bunk and look out. Magnificent weather. The wind, now coming from the east, has completely calmed. Surrounded by a beautiful forest of Canadian pines, there is no other sound of life apart from birds. To think, this land was once like this across its length and its breadth!

But we have no time to lose. Onward!

By sundown we are in Cheboygan and lie alongside our friends, the fishermen's boats. They stopped here where they have friends, also fishermen. We all go together for supper. On the way, we stop with a Polish family where we discover that there is a little Polish colony here, complete with a church. At my request, they take me in their Ford to visit the parish priest. In any case, we are not meeting our friends for supper until later in the evening.

So we set off; beyond the town there's a two-story wooden house. The upper floor is the parish school, downstairs is the church. We have seen this kind of church several times passing through the States. I knock on the door, a housekeeper opens it and leads me to the chancellery.

The parish priest, an overweight man of 40 or 50, is sitting at a desk.

"What do you want," he asks sharply.

I explain who we are and that we dropped in for a visit along the way.

"Haven't heard anything about you. Your pride is speaking for you. You know what the Lord Jesus said?"

"Forgive them, for they know not what they do, dear Father."

Our present port is not on the lake but on the Mackinac Straight that links three great lakes, the Huron, which we've just crossed, Superior to the north, and Michigan to our west. We have to sail almost the entire length of Lake Michigan to reach Chicago. The weather is absolutely fatal but we have no desire to sit around here, especially since we are but 20-30 km away from Mackinac Island, a great tourist centre just at the entrance to Lake Michigan. That's where wealthy Americans gather to escape the intense summer heat of Chicago. So next day at dawn...

We prepare for a short but uncomfortable voyage. Dressed in our oilskins and helmets, we firmly secure everything on the yacht. Off we go.

After leaving the port, we take the sheltered route along the first island, which makes the first part of our voyage relatively peaceful but later, when we are out in open water again, the conditions are frightful. The current from Superior is strong, as is the wind. The waves give the impression of doing perpendicular leaps, pounding the boat from all sides, flooding over the deck. Finally we arrive at the marina on the island.

All kinds of yachts are tied up here. It makes no sense to go out in this kind of weather. For the most part, these are large motorboats, and not really yachts. Some of these motorboats are of enormous proportions. Some are fifty metres long, and even more. The owners of the boats are millionaires from Chicago who have come here for a holiday. There are so many of them that there is no room for us. We are about to leave and go to a commercial basin when a marina employee spots us and gives us a signal to go to a small basin reserved for official guests. We moor in the designated spot and half an hour later, after tidying up the yacht we jump wearing spotless white trousers and caps, and navy blue blazers. We have practiced this and no matter what the conditions, we are capable of changing into our formal sailing wear with lightning speed. As we later find out, we were being observed from behind the curtains of the great boats whose occupants noted, and were duly impressed, by the rapid transformation of the barefoot, unkempt sailors into properly dressed yachtsmen.

We walk over to a rather elegant restaurant for breakfast, at least to have a look at the inmates but unfortunately the inmates must all be sleeping because there is no one there. When we return to the harbour it is almost noon. Wearing thick coats and scarves thrown over their beach pyjamas, bathing suits or tennis shorts, and despite the strong wind, the passengers of these luxurious boats are wandering around the pier. Seems our ensign is very intriguing.

The owners of these yachts had immediately ordered their captains to check their sea manuals and find out what country we belong to. Our club has the privilege of flying the military ensign with the added emblem of the club in its corner. The captains of the yachts on the Great Lakes, who have little experience with such things, were somewhat startled. Our appearance, therefore, spares them all the trouble of doing further research. We have to relate, for the thousandth time the story of our expedition. We are immediately issued invitations to every yacht and don't turn any down, being very keen to see the interiors of these boats—even if just briefly.

The Vice-Commodore of the Grosse Pointe Yacht Club, Mr. Jacobsen, invites us to dinner on his splendid yacht, and sends a tender to take us out. We are received with all the honours including cannon fire and the Polish ensign. Our host is a true fanatic of the sea. He rarely sets foot on land preferring to stay anchored in the harbour than being moored at the pier. He asks us about every technical detail of our voyage, all the time generously pouring rum.

He gives us a Danish sailing handbook as a souvenir, which we will later donate to the library of our club once we get back to Gdynia.

Our warmest friendship is established with the owners of a huge motor yacht, *Minoco*. Their burgee has a bi-coloured background on which are two sets of stars, four white ones and four blue, representing the four brother and four sisters of the Mills family who own the boat. After the death of their father, this extraordinarily close-knit family did everything together, both in their work and in their leisure. They own an enormous factory that makes machines and mechanical instruments of all sorts. Later on, we will frequently prowl around their factory because half of the workers are Poles.

Minoco, like all the other boats of this type, is luxuriously fitted with every possible kind of comfort.

Every cabin has its own bathroom and each one is decorated in a different colour. The boat also has its indispensible bar, a smoking room, a salon, a dining room and a gym.

We are welcome guests on every yacht and I don't know how long our stay might have gone on had we—Świechowski and

I—not got into an argument one night on the subject of our departure.

In the course of a heated conversation these words were uttered: "As far as I'm concerned, I could leave immediately," and that provoked the next statement: "Terrific, so that settles it, we are leaving right now!" With us there's not much space between words and actions.

So dressed as we are for an elegant evening, we jump onto our deck and without saying good-bye to anyone we cast off.

I can only guess at our hosts' surprise when they discover, next day, that we are gone.

Once we are beyond the breakwater, the weather turns ominous. Świechowski starts some breakneck tricks trying to get through the seas overrunning our deck every minute to get at our navigation lights to light them.

Holding the tiller between my knees I somehow manage to get my oilskins on. That done, all I have to do is figure out where we are to go, reading a soggy, wrinkled chart under the pale illumination of a flashlight.

In the morning I wake up my companion and turn the helm over to him. I have a pleasant surprise for him; we have several large rocks already behind us, the difficult passage through the isthmus is behind us, and we are now sailing down Lake Michigan in good weather, on calm water, and with full sails.

Suddenly we see a large boat speeding towards us. We recognize the uniformed crew of the coast guard. They order us to stop. We are quite taken aback: what's happening? But we obey the order at once.

As it turns out, during the most recent storm a yacht with a crew of three disappeared, its size and style matched ours, so they assumed we were it.

It was a week before the waves wash up on shore some rubber mattresses and a lifebelt from that yacht.

Apparently Lake Michigan never returns its victims.

We sail into Frankfurt at midnight.

Knowing that we would be getting there after dark, we stop at dinnertime at a coast guard station where the captain very kindly gives us a handmade sketch of the port.

Guided by this sketch we turn to port directly into the outer basin just as was suggested and then... we are grounded in the shoal. Neither motor nor the oars are of any help; we sit there with our keel in the mud unable to move forward or back. We spend the night in that position.

At dawn some fishermen are passing by. In response to our entreaties they throw us a line and try to pull us out, but the line snaps. They wave to us and go on their way. They have no time for these games.

Our situation is very unpleasant. The little port is deserted and there is no way we can get ashore to ask for help. We are sighted an hour later. A coast guard rescue boat arrives and within minutes pulls us out of the shoal. We moor in a convenient spot and head off to the local representative of Standard Oil, who immediately attends to everything.

The hole in the hull from a few days ago is repaired, the gas supply is refilled, although we really don't need it, but since we have an engine, why not use it at least when entering and leaving a port?

Since we have nothing to do while these things are being attended to, it occurs to us that we should walk over to the town hall, a little distance away. These are business hours, so why not a visit to the mayor?

A delightful secretary greets us in the office and informs us that the mayor is not in "but," she adds with a smile, "if you gentlemen are going to be around for the next couple of hours, I will show you the highlights of our town."

A perky little hat at an angle over her right eye, some powder on her turned-up nose, keys to the office turned over to the janitor, and we are off touring the little town and the surrounding area, which is amazingly beautiful. She draws our attention to the new golf club on a hill overlooking the lake and surrounded by forest.

When the time for our departure comes, I regretfully take my leave to attend to a few details concerning our yacht assigning to my colleague the responsibility to escort our charming guide back to her office.

A couple of hours later we are sailing south with a brisk wind behind us. At sundown we sail into the marina at Ludington.

Our original plan was to cross to the other side of the lake so that we could get to see Milwaukee, another large centre of Polish Americans, before continuing to Chicago. However, we decide to leave that until the end of our voyage because Bob Rohn, one of the co-owners of the *Minoco*, invited us to the regatta at Ludington.

By pure luck, we moor exactly next to his racing boat so it is a happy surprise when we meet up with Bob next morning. The *Minoco* had brought them here during the night, while we slept. Bob and his wife disembarked and the *Minoco* continued south to Chicago. Their poor captain has to soon weave his way back because he received orders to take *Minoco* through the canals to Florida, where the family would spend the season.

Meanwhile, in his hometown, Bob treats us with gracious hospitality.

Bob is a philosopher, despite his young age. He much prefers the peaceful life in his small town to the hustle and bustle of a large city. We spend almost our whole time with him, except for one evening when we are invited for supper in the home of a Polish butcher who, upon hearing of our arrival, rushed to the marina to greet us.

The weather is not favourable for the regatta. The races will take place not out in the open on the lake but on an internal waterway[14].

We are not familiar with local practices, nor have we ever been on Bob's yacht but there's nothing one won't do for sport. We are up for the race.

There are about 20 yachts participating in the regatta. We sail up and back together, waiting for the race to begin. There are people from the town and surrounding areas all along the shore to watch. The wind is very strong but the water, on the inner waterway, is calm. The race starts in three minutes... in two... in one... and there's the shot! I am trying to recall the fine points of regattas on the Vistula. We are ahead of all the others except for one yacht that keeps getting in our way so as to prevent us from passing her. The course requires two turns around the reed. The first time around we notice that in a certain place where the coastline is steep there is very little wind. So on our second round we turn slightly towards that point. Our competitor also turns in our direction and before long he finds himself in the windless zone. As soon as we see his sails collapsing, we sharply turn in the other direction avoiding the quiet zone and passing by the competing yacht.

To be honest, we must admit that winning this race was largely due to the first rate quality of the yacht. Public opinion decides otherwise. Everybody knows, so it is said, that Bob Rohn, a very rich guy, had engaged a couple of professionals. So he won.

The next evening we are in Grand Haven, the home of our fishermen friends. From the first moment, we are at home. We spend the whole day with them. The shore here is a large sand dune with a few pine trees growing on it. To encourage tourism, the local government constructed several paved access roads. It costs nothing to stop and camp here. So as soon as it's warm and sunny, people arrive in their cars and camp under the sky. There are hundreds of cars parked along the stretch of the sands.

People bring all kinds of camping gear. Multi-coloured tents, some conical, some just a roof attached to the car, but for the most part, we see "trailers."

These trailers are real little houses on wheels built with very light materials; attached to the car, the average American can tour the entire United States bringing his bedroom, bathroom and kitchen with him, along with, of course, the lady of the house.

During our stay in Grand Haven we get initiated into all the details of a fisherman's profession. We learn how nets are cast, how they are repaired, what kinds of fish are in the lakes, what their market prices are, and thousands of other little details.

Before our departure our fisherman friend, holding on to our hands, says, "Listen boys, our lives are not easy here but if there's ever a need, if at some point you experience difficulties in your life, come back to us. You are true men of the sea, there will always be room for you here with us."

Good man, dear friend, I can't complain about the life fate dealt me but your words often come to mind.

We are now by the bascule bridge in Michigan City, just 60 km away from our goal: Chicago. The chairman of the welcoming committee in Chicago is the director of the Customs office at the port and head of the coast guard in the region, Mr. Antoni Czarnecki.

I have his phone number and call him right away. I expect to hear a deep voice and words of great import as befitting his exalted position and as our guardian, host and organizer, but instead his voice is full of excitement and pleasure welcoming us wholeheartedly.

"Well, well, well, we are delighted that you are so close by, almost here. That's terrific! You know that we had lost sight of you for a couple of days. I'm issuing orders right away for the Coast Guard ship for Chicago to meet you and provide an honorary escort; Together with other dignitaries, I will be there for the salute by cannon fire when you arrive."

We are awakened during the night by a strong swell pushing us against the pier. We had secured the hull against this by making fenders of old tires that we had found but that didn't help much. At the first sign of dawn we ask to have the bridge raised to let us sail out into the river until we find a calm, safe place to anchor. The wind howls mercilessly, thrashing the trees along the shore, bending them almost to the breaking point.

A few hours later we are heading into a local Coast Guard port to see the Captain, with whom we walk along the pier, because getting to the breakwater is out of the question. The howling wind is deafening, to speak to one another we have to shout directly in each other's ears as we curse the weather. The worst thing is that it's blowing straight from the lake hitting the breakwater and throwing the water several metres straight up into the air or right over the top of the breakwater.

At the entrance to the port it is worse still because, with no barrier to stop them the waves fall into the harbour, churning and tossing walls of water around, some of them reaching all the way to the lighthouse beside the harbour.

"It's impossible to leave in these conditions," I tell the captain as we are walking back.

"That's what I think too," he replied. "That's what I told Mr. Czarnecki this morning on the phone, but he brushed me off telling me I knew nothing about sailing, that in Poland you do this sort of thing every day, that if you managed to cross the ocean you'll manage to sail across Lake Michigan and at nine o'clock he is leaving with the entire committee to meet you."

There is a moment of consternation, but really, there is no choice… we have to go.

We get back to our yacht and begin preparations without saying a word. The sails are reefed, everything is gauged, every opening is covered with tarps; we put on our yellow oilskins and using our engine we sail out of the river to the port.

The situation is delicate. The wind is irregular, we can't move properly, and the waves are forcing us against rocks or the breakwater. So we decide to leave the port not through the main access but by the side, by the "kitchen door," between the end of the breakwater and the shallows next to the beach where there is a channel a few metres wide and a couple of metres deep.

We go as fast as we can along the length of the breakwater and, as soon as we get to the end, we make a sharp turn, push all the sails to the other side, and out we go. My heart stops for what seems like an eternity as the waves hit us with a fury trying to force us back and crush us against the breakwater but our sails are now in the position we want with the wind and we're doing well. Gradually we are getting away from the danger and out into the open, where we feel at home. The waves are long, so that the yacht goes into them and out again very gracefully. After a while a rescue boat from the Coast Guard joins us to provide an honorary escort. It's a large, 30-metre boat with a 100-horsepoweer engine, with a six-man crew. Although we are only about 40 metres apart the huge waves often prevent us from seeing one another. A couple of hours later we see a long, grey ship. That's Mr. Czarnecki's "gunboat," with naval artillery mounted on its 120-metre long deck like a real warship.

We salute with our flag and wave to one another, though that's not easy because the waves prevent us from seeing one another, and then, over the top of the high seas we can see the tops of signalling masts.

As often happens on the Great Lakes the wind dies down suddenly, the waves calm down. We sail closer to the ship.

At this moment Świechowski yells out: "We can see it now!"

I look ahead... far on the horizon... there are matchboxes of different heights, standing one next to another. The land itself is not yet visible. Those buildings must be of a monstrous size to be visible from such a great distance.

Chicago! That huge powerhouse of a city, the second largest centre of Polish people after Warsaw.

We sat silently, our eyes glued on the buildings in the distance. It's hard, really hard, to express in words what I am feeling at this moment. We had set ourselves a goal, and we reached it. We can now say to one another: We did it!

It's impossible to describe what we have experienced, what torment we endured. Well actually, at this moment, everything is forgotten, nothing matters. Neither what was, nor what will be.

Now even the elements concede victory to us. The wind has completely died down, the waves are gone. A large vessel is coming out to greet us.

Arriving in Chicago

As we enter the port, we steer towards the small island of the coast guard where we will spend the night. A few minutes later the welcoming delegation joins us, accompanied by reporters.

We shake hands, we talk a lot, but our mind stubbornly returns to just one thought: We made it to Chicago.

Endnotes

13. Katowice is the capital of the Silesia region.
14. Possibly Père Marquette.

Epilogue

The next day, August 25, with great pomp and ceremony, we sailed *Dal* to the "lagoon" at the Chicago World's Fair. We had our own little harbour reserved for us along the main exhibition promenade, not far from Admiral Byrd's polar ship. A stand showing charts and maps of our voyage is erected nearby while a kiosk exhibits more maps and information about the port at Gdynia as well as about our adventure. Promoting Poland, after all, is the objective. So a tour of the boat itself is part of the exhibit, and all of it free, of course.

August 26 is "Polish Day," and on this day there are over 200,000 visitors to the exhibition, the highest attendance on record.

At the events on that day and some following days, we stand on a specially made podium addressing endless crowds using a megaphone.

A few weeks later, the organizers of the exhibition

Dal on exhibit at World's Fair.

arrange a great event at which they award commemorative medals to us, the same as the ones presented to the members of Admiral Byrd's Polar Expedition, and to the squadron of planes from Italian General Italo Balbo's transatlantic flight.

Along the way there are events at one club after another, both Polish and American.

Unfortunately, as time passed our funds diminished; time to plan our return home. On the other hand, conscience does not permit removing our yacht from the exhibition grounds before the end of the Fair because, unlike other countries, Poland has no presence apart from *Dal*.

But, the Polish community in Chicago comes up with an idea. There's a park in Chicago where a copy of the ship that brought Columbus to America—the Santa Maria—is on display, a gift from the state of Italy. In another place, Norway is represented by a model of a Viking ship like the one Erik the Black sailed to Labrador[1]. Clearly, this calls for a boat from Poland to be in Chicago as well, "a monument," complete with a bronze plaque inscribed with the essential details.

The committee sponsoring us decides therefore to acquire *Dal*. The parks department of the city even offers a place in a very popular park, the biggest municipal park with 41 streets leading to it. And, although the Chicago Poles have no money, they are determined to raise the funds for this goal. After much thought, we come to this decision: Świechowski will return to Poland because it's expensive to keep both of us in Chicago and the money we had for our expedition is running very low. I, on the other hand, stay behind in the company of the indomitable Mr. Czarnecki and take on a series of unpaid public appearances giving lectures about Poland, Polish sailing, and our own transatlantic voyage. My talks are mostly in Polish, though on occasion in English too. I have a few engagements to speak in French as well. I eventually reach what I consider a fantastic number of lectures: eighty.

Having been persuaded that this is a noble cause, it all goes well.

At the end of November, when the exhibition closes, the committee holds a meeting and informs me that despite their most energetic efforts they managed to raise only $1,500.00 and request that I accept this sum as payment for all my lectures and to cover any expenses I have incurred.

The yacht had cost me over 10,000 złoty; could I, after all this, sell it to them for this amount?

Well, it has to be my grand gesture. I turn over *Dal*'s documents to the committee and sign an agreement of donation.

I am then on my own. Thanks to a great many friends I am able to take advantage of being in the US to travel the length and breadth of the country, all the way to California and across to New Orleans and Florida, in short, from Canada to the Mexican border.

I return to New York in February. The ever-gracious gentlemen of the Gdynia-Amsterdam Line offer me first class passage to Poland, as they had earlier done for my friend, Świechowski.

Upon my return, at a celebration in our honour at the Officers' Yacht Club of the Republic of Poland, Świechowski and I are decorated with the "Cross of Merit." Unfortunately, our other crew member, Witkowski, is not with us.

We are both very touched by all this. As for me, despite all the lectures I have given in America, I am totally incapable of uttering anything sensible. My tongue got so stuck in my throat that Świechowski, who is usually so indulgent of my inept orations, must have been completely embarrassed on my behalf.

Ah well, as they say, "Nobody is a prophet in his own country."

And now, what next?

Nothing. We go back to work.

Endnote

1. No doubt referring to Leif Erikson, Erik the Red's son.

Dal Dimensions

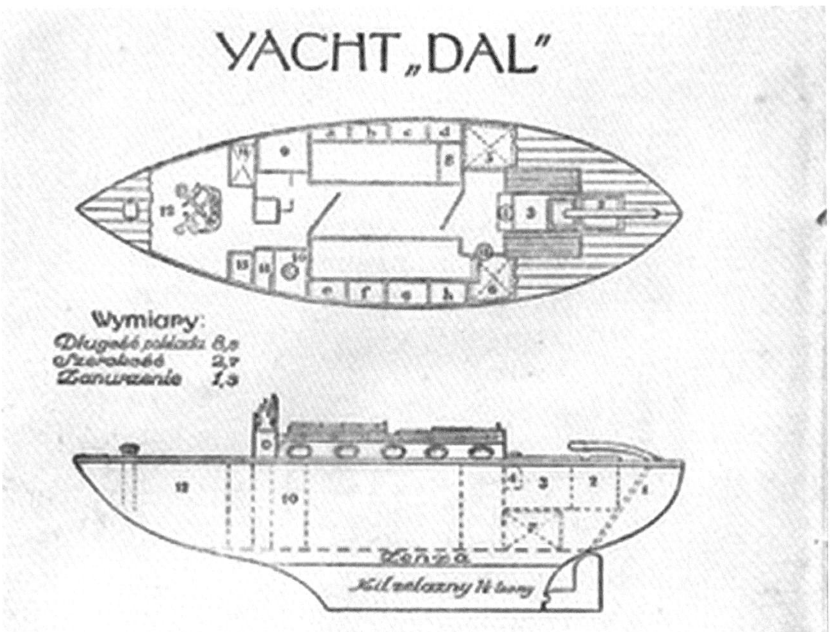

Top image:

LOA (Length Over All)	8.5
Beam	2.7
Draft	1.3

Bottom image

Zenza	Bilge
Kil zelazny	Iron keel 1.5 tons

Objasnienia—layout

1. Afterpik (konserwy)	Afterpeak (cans, lines, grits, pasta)
2. Skrzynia botsmanska	Bosun's chest (tools)
3. Kokpit	Cockpit with side benches
4. Kompas	Compass
5. Koja Witkowskiego	Witkowski's berth, tank underneath
6. Zagiel-koja	Sail bunk, tank underneath
7. Duzy tank	Large tank under the cockpit
8. Radio odbiorcze	Radio receiver
a,b,c,d	Storage cabinets
9. Szafa na ubranie	Wardrobe locker
10. Kuchnia	Galley
11. Przedni tank	Forward tank
12. Forpik	Forepeak (anchors, lines, chains, kerosene cans, etc.)
13. Pompa	Pump
14,15. Szafki	Cabinets

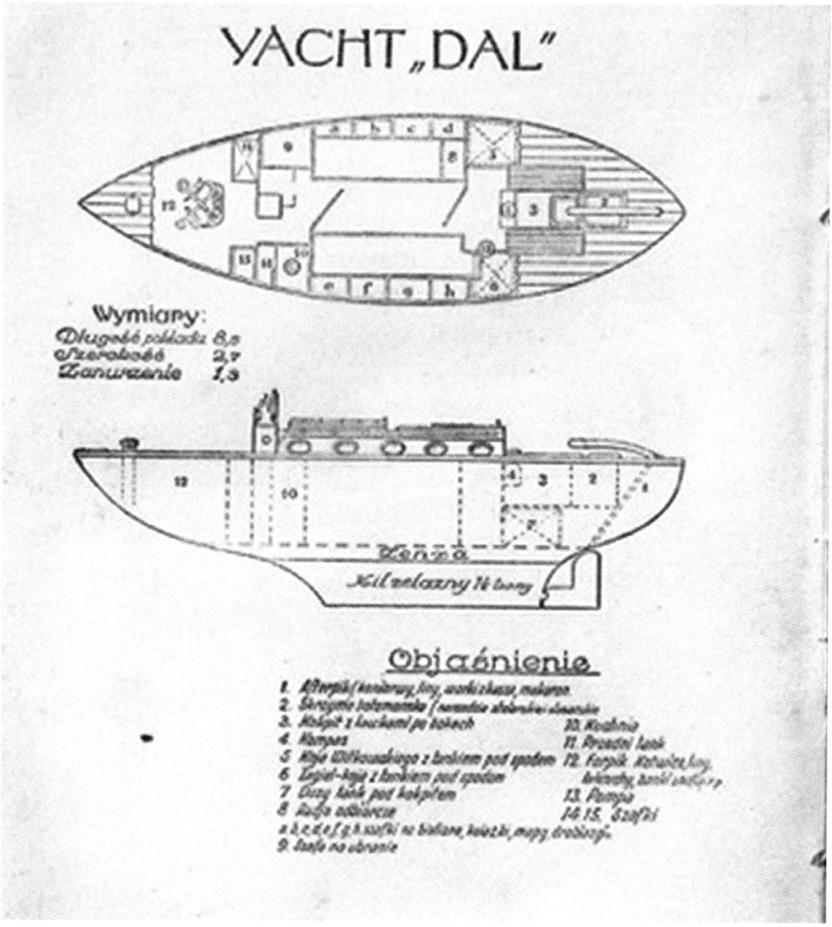

Top left:

Basic Rig

Gafel	Gaff
Fok	Jib
Grot	Mainsail

Bottom left:

Beam reach/close-hauled, moderate wind

Kliwer	Headsail or Foresail
Gafel	Gaff
Fok	Jib
Grot	Mainsail

Top right:

Fok Szt.	Storm jib
Grot sztormowy	Trysail

Bottom right:

Running/reaching, moderate wind

Gafel	Gaff
Fok	Jib
Grot	Mainsail
Balon	Spinnaker
Wystawiony spinaker-bom	Spinnaker pole
Saling	Spreader

Centre: N.B. For clarity all rigging (forestays, halyards, sheets, etc.) are not shown.

Sources

Anna and Andrzej Bohomolec Collection, Piłsudski Institute, NY.
Anthony Keith, Ottawa. Letters, photographs, recollections.
Maria Skarżynska, Calgary. Family correspondence, recollections.
Antonella Bell, Edmonton, Personal papers of Andrzej Bohomolec.

For general background:

Davies, Norman, *Vanished Kingdoms: The History of Half-Forgotten Europe*, Penguin Books, 2011

Davies, Norman, *God's Playground, Vol. I*, Oxford: Oxford University Press, 2001

Halecki, Oskar, *Borderlands of Western Civilization*, The Ronald Press, New York, 1952

Snyder, Timothy, *The Reconstruction of Nations: Poland, Ukraine, Lithuania, Belarus, 1569-1999*, New Haven & London: Yale University Press, 2003

For Bohomolec/Bogoria lineage & history:

Beyzym, Jan & Bilinski Jerzy, *Polski Słownik Biograficzny*, Vol. II/1, Krakow: Nakładem Polskiej Akademiji Umiejetności, 1936

Boniecki, Adam, *Herbarz Polski, Część I: Wiadomości Historiyczno-Genealogiczne o Rodach Szlacheckich*, Warszawa: Gebethner & Wolff, 1899

Ciechanowicz, Jan, *Rody rycierskie Wielkiego Księstwa Litewskiego, Tom II*, Rzeszów: Biblioteka Stowarzyszenia Dom Polski SARMACJA, 2001

Kulikowski, Andrzej, *Wielki Herbarz Rodów Polskich*

Pizunski, Paweł, *Wielki Leksykon Rycerstwa Polskiego*, Gdansk: L & L: 2007

Siwik, Jan, *Encyklopedia Nazwisk I Przydomków Szlacheckich*, Warszawa: Wydawnictwo Kastor, 2010

Szymański, Józef, *Herbarz Średniowiecznego Rycerstwa Polskiego*, Warszawa: Wydawnictwo Naukowe PWN

Illustrations:

Pen & ink drawings by Adam Werka, courtesy of Dr. Joanna Werka, president of the Adam Werka Foundation in Warsaw.

Family photographs courtesy of Anthony Keith.

Photograph of Rozentowo from *Dzieje rezydencji na dawnych kresach Rzeczpospolitej*, in Wrocław, Poland. Courtesy Jan Dowgiałło.

Rozentowo cheese ad: http://warszawskikram.blogspot.com/2013/11/august-szmolke.html accessed March 10, 2019.

Acknowledgements

Every book has godparents. Top of the list are Anthony Keith and Iola Price. Anthony introduced me to his stepfather, André Bohomolec. Anthony and Iola provided many enjoyable afternoons of tea and conversation along with Iola's expert and time-consuming scanning of old photographs and Anthony's patient reading of endlessly revised texts. Their encouragement and support underpinned the whole project.

Paweł Wołowski, 1st Secretary, Embassy of Poland in Ottawa, introduced me to Anthony Keith. Paweł was always ready to help and his enthusiasm for the project never wavered.

My dear friend in Warsaw, Małgorzata Dzieduszycka, who can find everyone and everything in her fair city and helped me understand some tricky Polish phrases. Globe-circling sailor Krzysztof Baranowski explained technical terms and the mysteries of the ocean's winds and currents, and he has known about André ever since he started sailing.

Maria Skarżyńska, in Calgary, for sharing her memories and family correspondence.

With thanks to Iwona Skorga and Jolanta Szczepkowska of the Piłsudski Institute in New York; the late Dr. Thaddeus Radzilowski, Detroit; Elżbieta Kieszczynska, NY; Andrew Ostrowski, Ottawa; Pierre Gratton, Ottawa; and Richard Gratton, Hudson, Quebec.

Everyone was very helpful but if there are mistakes, they are mine.

Irene

About Irene Tomaszewski

Irene Tomaszewski is a Canadian writer, translator and editor. Her books include "Codename Żegota: Rescuing Jews in Occupied Poland, 1942-1945 – The Most Dangerous Conspiracy in Wartime Europe," published in 2010 by Praeger, an imprint of ABC-CLIO, LLC, California, co-authored with Tecia Werbowski, and "Inside a Gestapo Prison: The Letters of Krystyna Wituska, 1942-44," published in 2006 by Wayne State University Press, Detroit MI. She lives in Ottawa, Canada.

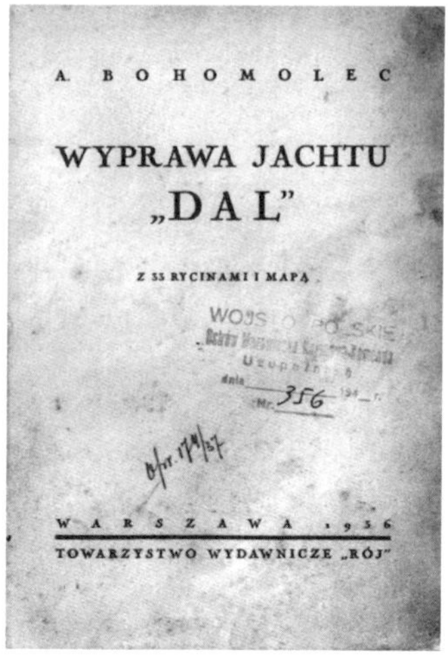

1st edition: Wyprawa Jachtu "Dal"
published by Towarzystwa Wydawnicze "RÓJ" 1936

2nd edition: Wyprawa Jachtu "Dal"
published by Wiedza Powszechna 1957